David Ca
Berkeley
August

THE MATHEMATICAL MANUSCRIPTS OF KARL MARX

Mathematical Manuscripts

of

KARL MARX

Translated by
C. Aronson and M. Meo

NEW PARK PUBLICATIONS

Published by New Park Publications Ltd.,
21b Old Town, Clapham, London SW4 0JT

The Mathematical Manuscripts
first published in German and Russian as
Karl Marx, *Mathematicheskie Rukopsii*
Nauka Press, Moscow, 1968

Set up, Printed and Bound by Trade Union Labour

Distributed in the United States by:
Labor Publications Inc.,
GPO 1876 NY
New York 10001

Printed in Great Britain by Astmoor Litho Ltd (TU),
21-22 Arkwright Road, Astmoor, Runcorn, Cheshire

ISBN 0 86151 029 1

Contents

PUBLISHERS' NOTE

The major part of this volume has been translated from Karl Marx, *Mathematicheskie Rukopsii*, edited by Professor S.A. Yanovskaya, Moscow 1968 (referred to in this volume as Yanovskaya, 1968). This contained the first publication of Marx's mathematical writings in their original form, alongside Russian translation. (Russian translation of parts of these manuscripts had appeared in 1933.) We have included the first English translation of Part I of the Russian edition, comprising the more or less finished manuscripts left by Marx on the differential calculus, and earlier drafts of these. We have not translated Part II of the 1968 volume, which consisted of extracts from and comments on the mathematical books which Marx had studied. Professor Yanovskaya, who had worked on these manuscripts since 1930, died just before the book appeared. We include a translation of her preface, together with six Appendices, and Notes to Part I.

In addition, we include a review of Yanovskaya, 1968, translated from the Russian, by the Soviet mathematician E. Kol'man, who died in Sweden in 1979, and who had also been associated with these manuscripts since their first transcription.

The translation is by C. Aronson and M. Meo.

S.A. Yanovskaya

PREFACE TO THE 1968 RUSSIAN EDITION

Engels, in his introduction to the second edition of *Anti-Dühring*, revealed that among the manuscripts which he inherited from Marx were some of mathematical content, to which Engels attached great importance and intended to publish later. Photocopies of these manuscripts (nearly 1,000 sheets) are kept in the archives of the Marx-Lenin Institute of the Central Committee of the Communist Party of the Soviet Union. In 1933, fifty years after the death of Marx, parts of these manuscripts, including Marx's reflections on the essentials of the differential calculus, which he had summarised for Engels in 1881 in two manuscripts accompanied by preparatory material, were published in Russian translation, the first in the journal *Under the Banner of Marxism* (1933, no.1, pp.15-73) and the second in the collection *Marxism and Science* (1933, pp.5-61). However, even these parts of the mathematical manuscripts have not been published in the original languages until now.

In the present edition all of the mathematical manuscripts of Marx having a more or less finished character or containing his own observations on the concepts of the calculus or other mathematical questions, are published in full.

Marx's mathematical manuscripts are of several varieties; some of them represent his own work in the differential calculus, its nature and history, while others contain outlines and annotations of books which Marx used. This volume is divided, accordingly, into two parts. Marx's original works appear in the first part, while in the second are found full expository outlines and passages of mathematical content.* Both Marx's own writings and his observations located in the surveys are published in the original language and in Russian translation.

* This volume contains a translation of the first part only.

Although Marx's own work, naturally, is separated from the outlines and long passages quoting the works of others, a full understanding of Marx's thought requires frequent acquaintance with his surveys of the literature. Only from the entire book, therefore, can a true presentation of the contents of Marx's mathematical writings be made complete.

Marx developed his interest in mathematics in connection with his work on *Capital*. In his letter to Engels dated January 11, 1858, Marx writes:

> 'I am so damnedly held up by mistakes in calculation in the working out of the economic principles that out of despair I intend to master algebra promptly. Arithmetic remains foreign to me. But I am again shooting my way rapidly along the algebraic route.' (K.Marx to F.Engels, *Works*, Vol.29, Berlin, 1963, p.256.)

Traces of Marx's first studies in mathematics are scattered in passages in his first notebooks on political economy. Some algebraic expositions had already appeared in notebooks, principally those dated 1846. It does not follow, however, that they could not have been done on loose notebook sheets at a much later time. Some sketches of elementary geometry and several algebraic expositions on series and logarithms can be found in notebooks containing preparatory material for *Critique of Political Economy* dating from April-June 1858.

In this period, however, the mathematical ideas of Marx proceeded only by fits and starts, mostly when he was not occupied with anything else. Thus on November 23, 1860 Marx wrote to Engels: 'For me to write is almost "out of the question". Mathematics is the single subject for which I still have the necessary "quietness of mind".' (Marx-Engels, *Works*, Vol.30, Berlin, 1964, p.113) In spite of this he invariably went on with his mathematical ideas, and already on July 6 1863 he wrote to Engels:

> 'In my free time I do differential and integral calculus. A propos! I have a surplus of books and will send one to you if you want to study this topic. I deem it almost indispensable for your military studies. By the way, it is a much easier part of mathematics (involving mere technique) than the higher parts of algebra, for instance. Outside of knowledge of the usual algebra and trigonometry there is nothing else necessary to study, except for general familiarity with the conic sections.' (*Ibid.*, p.362)

Also, in the appendix to an unpreserved letter from the end of 1865 or beginning of 1866 Marx explained to Engels the essentials of the differential calculus in an example of the problem of the tangent to the parabola.

However, he was still concerned first of all with the basics of mathematics in their connection with political economy. Thus in 1869, in relation to his studies of questions of the circulation of capital and the role of promissory notes in inter-governmental calculations, Marx familiarised himself with the long course of commercial arithmetic, Feller and Odermann, which he outlined in detail (cf. mss.2388 and 2400). It was characteristic of Marx's survey techniques that, coming across some question of which he did not already feel himself in command, Marx was not content until he had mastered it completely, down to its foundations. Every time Feller and Odermann used some mathematical technique, Marx considered it necessary to re-commit it to memory, even if it was known to him. In his surveys of commercial arithmetic — these and also much later ones, cf. mss.3881, 3888, 3981 — are found insertions, moreover, of purely mathematical content in which Marx advanced even further into fields of higher mathematics.

In the 1870s, starting in 1878, Marx's thoughts on mathematics acquired a more systematic character. Concerning this period Engels in the introduction to the second edition of *Capital*:

> 'After 1870 came another pause caused mainly by the painful illnesses of Marx. By habit, he usually filled his time studying; agronomy, American and especially Russian land relationships, monetary markets and banks, and finally natural science: geology and physiology, and particularly his own mathematical work, all go to make up the contents of numerous notebooks from this period.' (Marx-Engels, *Works*, Vol.24, Berlin 1963, p.11)

At the same time the problems of applying mathematics to political economy continued to interest Marx. Thus in a letter to Engels of May 31, 1873 Marx wrote:

> 'I have just sent Moore a history which *privatim* had to be smuggled in. But he thinks that the question is unsolvable or at least *pro tempore* unsolvable in view of the many parts in which facts are still to be discovered relating to this question. The matter is as follows: you know tables in which prices, calculated by percent

etc., etc. are represented in their growth in the course of a year etc. showing the increases and decreases by zig-zag lines. I have repeatedly attempted, for the analysis of crises, to compute these "ups and downs" as fictional curves, and I thought (and even now I still think this possible with sufficient empirical material) to infer mathematically from this an important law of crises. Moore, as I already said, considers the problem rather impractical, and I have decided for the time being to give it up.' (Marx-Engels, *Works*, Vol.33, Berlin, 1966, p.82).

Thus it is clear that Marx was consciously leading up to the possibility of applying mathematics to political economy. Given the full texts of all Marx's mathematical manuscripts in the second part of our book, it still does not fully answer the question of what impelled Marx to proceed to the differential calculus from the study of algebra and commercial arithmetic. Indeed the mathematical manuscripts of Marx begin precisely in this period when Marx was concerned with elementary mathematics only in connection with problems arising from his study of differential calculus. His studies of trigonometry and the conic sections are found exactly in this context, which he suggested to Engels to be indispensable.

In differential calculus, however, there were difficulties, especially in its fundamentals — the methodological basis on which it was built. Much light was thrown on this condition in Engels's *Anti-Dühring*.

> 'With the introduction of variable magnitudes and the extension of their variability to the infinitely small and infinitely large, mathematics, in other respects so strictly moral, fell from grace; it ate of the tree of knowledge, which opened up to it a career of most colossal achievements, but at the same time a path of error. The virgin state of absolute validity and irrefutable certainty of everything mathematical was gone forever; mathematics entered the realm of controversy, and we have reached the point where most people differentiate and integrate not only because they understand what they are doing but from pure faith, because up to now it has always come out right.' (*Anti-Dühring*, p.107)

Naturally Marx was not reconciled to this. To use his own words, we may say that 'here, as everywhere' it was important for him 'to tear off the veil of mystery in science'. (see p.109) This was of the more importance, since the procedure of going from elementary mathematics to the mathematics of a variable quantity must be of an essentially dialectical character, and Marx and Engels considered themselves obliged to show how to reconcile the materialist dialectic not only with the social sciences, but also with the natural sciences and mathematics. The examination by dialectical means of mathematics of variable quantities may be accomplished only by fully investigating that which constitutes 'a veil surrounded already in our time by quantities, which are used for calculating the infinitely small — the differentials and infinitely small quantities of various orders'. (Marx-Engels, *Works*, Vol.20, Berlin, 1962, p.30) Marx placed before himself exactly this problem, the elucidation of the dialectic of symbolic calculation, operating on values of the differential.

Marx thought about mathematics independently. The only person to whom he turned was his friend Samuel Moore, whose understanding of mathematics was at times rather limited. Moore could not render any essential help to Marx. Moreover, as can be observed in remarks that Moore made concerning the 1881 manuscripts (which Marx sent Engels) containing Marx's expository ideas on the derivation and meaning of the symbolic differential calculus, Moore simply did not understand these ideas. (cf. Marx's letter to Engels, this volume p.xxx)

Marx studied textbooks of differential calculus. He oriented himself with books used at courses in Cambridge University, where in the 17th century Newton held a chair of higher mathematics, the traditions of which were kept by the English up to Marx's day. Indeed, there was a sharp struggle in the 20s and 30s of the last century between young English scholars, grouped about the 'Analytical Society' of mathematicians, and the opposing established and obsolete traditions, converted into untouchable 'clerical' dogma, represented by Newton. The latter applied the synthetic methods of his *Principia* with the stipulation that each problem had to be solved from the beginning without converting it into a more general problem which could then be solved with the apparatus of calculus.

In this regard, the facts are sufficiently clear that Marx began studying differential calculus with the work of the French abbot Sauri, *Cours complet de mathématiques* (1778), based on the methods of Leibnitz and written in his notation, and that he turned next to the *De analyse per aequationes numero terminorum infinitas* of Newton (cf.ms.2763). Marx was so taken with Sauri's use of the Leibnitzian algorithmic methods of differentiation that he sent an explanation of it (with application to the problem of the tangent to the parabola) in a special appendix to one of his letters to Engels.

Marx, however, did not limit himself to Sauri's *Cours*. The next text to which he turned was the English translation of a modern (1827) French textbook, J.-L. Boucharlat's *Eléments de calcul différentiel et du calcul intégral*. Written in an eclectic spirit, it combined the ideas of d'Alembert and Lagrange. It went through eight editions in France alone and was translated into foreign languages (including Russian); the textbook, however, did not satisfy Marx, and he next turned to a series of monographs and survey-course books. Besides the classic works of Euler and MacLaurin (who popularised Newton) there were the university textbooks of Lacroix, Hind, Hemming and others. Marx made scattered outlines and notations from all these books.

In these volumes Marx was interested primarily in the viewpoint of Lagrange, who attempted to cope with the characteristic difficulties of differential calculus and ways of converting calculus into an 'algebraic' form, i.e., without starting from the extremely vague Newtonian concepts of 'infinitely small' and 'limit'. A detailed acquaintance with the ideas of Lagrange convinced Marx, however, that these methods of solving the difficulties connected with the symbolic apparatus of differential calculus were insufficient. Marx then began to work out his own methods of explaining the nature of the calculus.

Possibly the arrangement of Marx's mathematical writings as is done in the second half of the volume permits a clarification of the way in which Marx came onto these methods. We see, for example, beginning with the attempt to correct Lagrange's outlook how Marx again turned to algebra with a complete understanding of the algebraic roots of the differential calculus. Naturally, his primary interest here was in the theorem of the multiple roots of an algebraic equation, the finding of which was closely connected with the successive differentiations of equations. This question was especially treated by Marx in the series of manuscripts 3932, 3933, appearing here under the titles 'Algebra I' and 'Algebra II'. Marx paid special attention to the important theorems of Taylor and MacLaurin. Thus arrived his manuscripts 3933, 4000, and 4001, which are impossible to regard simply as outlines and the texts of which are, therefore, given in full.

Generally speaking in the outlines Marx began more and more to use his own notation. In a number of places he used special notation for the concept of function and in places $\frac{dy}{dx}$ for $\frac{0}{0}$. These symbols are met passim a number of other manuscripts (cf. 2763, 3888, 3932, 4302).

Convinced that the 'pure algebraic' method of Lagrange did not solve the difficulties of the foundations of the differential calculus and already having his own ideas on the nature and methods of the calculus, Marx once again began to collect textual material on the various ways of differentiating (cf. mss. 4038 and 4040). Only after reading the expositions suggesting (for certain classes of functions) the methods of 'algebraically' differentiating, only after constructing sketches of the basic ideas did he express his point of view. These are exhibited here in the manuscripts and variants published in the first part of this volume. We now proceed to the contents of these manuscripts.

In the 1870s, from which date the overwhelming majority of Marx's mathematical works, contemporary classical analysis and characteristic theories of the real numbers and limits were established on the European continent (principally in the works of Weierstrass, Dedekind and Cantor).

This more precise work was unknown in the English universities at that time. Not without reason did the well-known English mathematician Hardy comment in his *Course of Pure Mathematics*, written significantly later (1917): 'It [this book] was written when analysis was neglected in Cambridge, and with an emphasis and enthusiasm which seem rather ridiculous now. If I were to rewrite it now I should not write (to use Prof. Littlewood's simile) like a "missionary talking to cannibals",' (preface to the 1937 edition). Hardy had to note as a special achievement the fact that in monographs in analysis 'even in England there is now [i.e., in 1937] no lack'.

It is not surprising therefore that Marx in his mathematical manuscripts may have been cut off from the more contemporary problems in mathematical analysis which were created at that time on the Continent. Nonetheless his ideas on the nature of symbolic differential calculus afford interest even now.

Differential calculus is characterised by its symbols and terminology, such notions as 'differential' and 'infinitely small' of different orders, such symbols as dx, dy, d^2y, d^3y . . . $\frac{dy}{dx}$, $\frac{d^2y}{dx^2}$, $\frac{d^3y}{dx^3}$ and others. In the middle of the last century many of the instructional books used by Marx associated these concepts and symbols with special methods of constructing quantities different from the usual mathematical numbers and functions. Indeed, mathematical analysis was obliged to operate with these special quantities. This is not true at the present time: there are no special symbols in contemporary analysis; yet the symbols and terminology have been preserved, and even appear to be quite suitable. How? How can this happen, if the corresponding concepts have no meaning? The mathematical manuscripts of Karl Marx provide the best answer to this question. Indeed, such an answer which permits the understanding of the essence of all symbolic calculus, whose general theory was only recently constructed in contemporary mathematical logic.

The heart of the matter is the operational role of symbols in the calculus. For example, if one particular method of calculation is to be employed repeatedly for the solution of a range of problems then the special symbol appropriately chosen for this method briefly designates its generation, or as Marx calls it, its 'strategy of action'. That symbol, which comes to stand for the process itself, as distinct from the symbolic designation introduced for the process, Marx called 'real'.

Why then introduce an appropriately chosen new symbol for this? Marx's answer consists in that this gives us the opportunity not to execute the entire process anew each time, but rather, using the fact of previously having executed it in several cases, to reduce the procedure in more complicated cases to the procedure of the more simple ones. For this it is only necessary, once the regularities of the particular method are well-known, to represent several general rules of operation with new symbols selected to accomplish this reduction. And with this step we obtain a calculus, operating with the new symbols, on its, as Marx called it, 'own ground'. And Marx thoroughly clarifies, by means of the dialectic of the 'inverted method', this transition to the symbolic calculus. The rules of calculus allow us on the other hand not to cross over from the 'real' process to the symbolic one but to look for the 'real' process corresponding to the symbol, to make of the symbol an operator — the above-mentioned 'strategy of action'.

Marx did all this in his two fundamental works written in 1881 and sent to Engels: 'On the concept of the derived function' (see p.3) and 'On the differential' (p.15). In the first work Marx considers the 'real' method, for several types of functions, to find the derived functions and differentials, and introduces appropriate symbols for this method (he calls it 'algebraic' differentiation). In the second work he obtains the 'inverted method' and transfers to the 'own ground' of differential calculus, employing for this aim first of all the theorem on the derivative of a product which permits the derivative of a product to be expressed as the sum of the derivatives of its factors. Employing his own words, 'thus the symbolic differential coefficient becomes the *autonomous starting point* whose real equivalent is first to be found . . . Thereby, however, the differential calculus appears as a specific type of calculus which already operates independently on its own ground (*Boden*). For its starting points $\frac{du}{dx}$, $\frac{dz}{dx}$, belong only to it and are mathematical quantities characteristic of it.' (pp.20-21). For this they 'are suddenly transformed into *operational symbols* (*Operationssymbole*), into symbols of the process which must be carried out . . . to find their "derivatives". Originally having arisen as the symbolic expression of the "derivative" and thus already finished, the symbolic differential coefficient now plays the role of the symbol of that operation of differentiation which is yet to be completed.' (pp.20-21).

In the teachings of Marx there were not yet the rigorous definitions of the fundamental concepts of mathematical analysis characteristic of contemporary mathematics. At first glance the contents of his manuscripts appear therefore to be archaic, not up to the requirements, say, of Lagrange, at the end of the 18th century. In actuality, the fundamental principle characteristic of the manuscripts of Marx has essential significance even in the present day. Marx was not acquainted with contemporary rigorous definitional concepts of real number, limit and continuity. But he obviously would not have been satisfied with the definitions, even if he had known them. The fact is Marx uses the 'real' method of the search for the derivative function, that is the algorithm, first, to answer the question whether there exists a derivative for a given function, and second, to find it, if it exists. As is well known, the concept of limit is not an algorithmic concept, and therefore such problems are only solvable for certain classes of functions. One class of functions, the class of algebraic functions, that is, functions composed of variables raised to any power, is represented by Marx as the object of 'algebraic' differentiation. In fact, Marx only deals with this sort of function. Nowadays the class of functions for which it is possible to answer both questions posed above has been

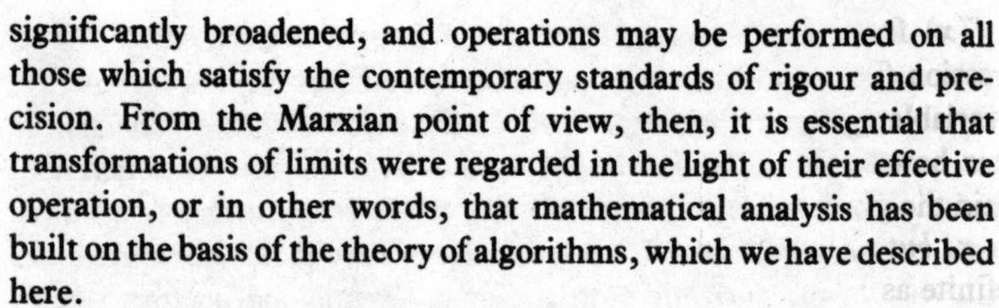

significantly broadened, and operations may be performed on all those which satisfy the contemporary standards of rigour and precision. From the Marxian point of view, then, it is essential that transformations of limits were regarded in the light of their effective operation, or in other words, that mathematical analysis has been built on the basis of the theory of algorithms, which we have described here.

We are certainly well acquainted with Engels's statement in the *Dialectics of Nature* that 'the turning point in mathematics was Descartes' introduction of *variable quantities*. Thanks to this *movement* came into mathematics and with it the *dialectic* and thanks to this *rapidly* became *necessary differential and integral calculus*, which arose simultaneously and which generally and on the whole were completed and not invented by Newton and Leibnitz' (*Dialectics of Nature* p.258).

But what is this 'variable quantity'? What is a 'variable' in mathematics in general? The eminent English philosopher Bertrand Russell says on this point, 'This, naturally , is one of the most difficult concepts to understand,' and the mathematician Karl Menger counts up to six completely different meanings of this concept. To elucidate the concept of variables — in other words, of functions — and that of variables in mathematics in general, the mathematical manuscripts of Marx now represent objects of essential importance. Marx directly posed to himself the question of the various meanings of the concepts of function: the functions 'of *x*' and functions 'in *x*' — and he especially dwelt on how to represent the mathematical operation of change of variables, in what consists this change. On this question of the means of representation of the change of variables Marx placed special emphasis, so much so that one talks characteristically of the 'algebraic' method of differentiation, which he introduced.

The fact is, Marx strenuously objected to the representation of any change in the value of the variable as the increase (or decrease) of previously prepared values of the increment (its absolute value). It seems a sufficient idealisation of the real change of the value of some quantity or other, to make the assertion that we can precisely ascertain *all* the values which this quantity receives in the course of the change. Since in actuality all such values can be found only approximately, those assumptions on which the differential calculus is based must be such that one does not need information about the entirety of values of any such variable for the complete expression of the derivative function $f'(x)$ from the given $f(x)$, but that it be sufficient to have the expression $f(x)$. For this it is only required to know that the value of the variable x changes actually in such a way that in a selected (no matter how small) neighbourhood of each value of the variable x (within the given range of its value) there exists a value x_1, different from x, *but no more than that*. 'x_1 therefore remains just exactly as indefinite as x is.' (p.88)

It stands to reason from this, that when x is changed into x_1, thereby generating the difference $x_1 - x$, designated as Δx, then the resulting x_1 becomes equal to $x + \Delta x$. Marx emphasised at this point that this occurs only *as a result* of the change of the value x into the value x_1 and does not precede this change, and that to represent this x_1 as known as the fixed expression $x + \Delta x$ carries with it a distorted assumption about the representation of movement (and of all sorts of change in general). Distorted because in this case here, 'although in $x + \Delta x$, Δx is equally as indeterminate in quantity as the undetermined variable x itself; Δx is determined separately from x, a distinct quantity, like the fruit of the mother's womb, with which she is pregnant.' (p.87)

In connection with this Marx now begins his determination of the derived function $f'(x)$ from the function $f(x)$ with the change of x into x_1. As a result of this $f(x)$ is changed into $f(x_1)$, and there arise both differences $x_1 - x$ and $f(x_1) - f(x)$, the first of which is obviously different from zero as long as $x_1 \neq x$.

'Here the increased x, is distinguished as x_1, *from itself*, before it grows, namely from x, but x_1 does not appear as an x increased by Δx, so x_1 therefore remains just exactly as indefinite as x is.' (p.88)

The real mystery of differential calculus, according to Marx, consists in that in order to evaluate the derived function at the point x (at which the derivative exists) it is not only necessary to go into the neighbourhood of the point, to the point x_1 different from x, and to form the ratio of the differences $f(x_1) - f(x)$ and $x_1 - x$ that is, the expression $\frac{f(x_1) - f(x)}{x_1 - x}$, but also to return again to the point x; and to return not without a detour, with special features relating to the concrete evaluation of the function $f(x)$, since simply setting $x_1 = x$ in the expression $\frac{f(x_1) - f(x)}{x_1 - x}$ turns it into $\frac{f(x) - f(x)}{x - x}$; that is, into $\frac{0}{0}$, or in other words into meaninglessness.

This character of the evaluation of the derivative, in which is formed the non-zero difference $x_1 - x$ and the subsequent — after the construction of the ratio $\frac{f(x_1) - f(x)}{x_1 - x}$ — dialectical 'removal' of this difference, is still preserved in the present-day evaluation of the derivative, where the removal of the difference $x_1 - x$ takes place with the help of the limit transition from x_1 to x.

In his work 'Appendix to the manuscript "On the history of the differential calculus", Analysis of the Method of d'Alembert' Marx also spoke of the 'derivative' essentially as the limit of the value of the ratio $\frac{f(x_1) - f(x)}{x_1 - x}$, although he denoted it with other terms. In fact the confusion surrounding the terms 'limit' and 'limit value', concerning which Marx observed, 'the concept of value at the limit is easily misunderstood and is constantly misunderstood', prompted him to replace the term 'limit' with 'the absolute minimal expression' in the determination of the derivative. But he did not insist on this replacement, however, foreseeing that the more precise definition of the concept of limit, with which he familiarised himself in Lacroix's long *Traité du calcul différentiel et du calcul intégral* — a text which satisfied Marx significantly more than others — could result further on in the introduction of unnecessary new terms. In fact Marx wrote of the concept of limit, 'this category which Lacroix in particular analytically broadened, only becomes important as a substitute for the category "minimal expression" ' (see p.68).

Thus Marx clarified the essentials of the dialectic connected with the evaluation of the derivative even in contemporary mathematical analysis. This dialectic, not a formal contradiction, makes, as will be shown below, the differential calculus of Newton and Leibnitz appear 'mystical'. To see this it is only necessary to recall that Marx by no means totally denied the representation of *any* change in the value of the variable as the addition of some 'increment' already having a value. On the contrary, when one speaks of the evaluation of *the result* of the already introduced change, one is induced to speak equally of the increase of the value of the variable (for example, of the dependence of the increase of the function on the increase in the independent variable), and 'the point of view of the sum' $x_1 = x + \Delta x$ or $x_1 = x + h$, as Marx calls it, becomes fully justified. To this transition from the 'algebraic' method to the 'differential' one Marx specially devoted himself in his last work 'Taylor's Theorem', which unfortunately remains unfinished and is therefore only partially reproduced in the first part of the present book. (A very detailed description of this manuscript of Marx, with almost all of the text, appears in the second part of the book, [Yanovskaya, 1968 pp.498-562]).

Here Marx emphasises that, while in the 'algebraic' method $x_1 - x$ consists solely for us as the form of a difference, and not as some $x_1 - x = h$ and therefore not as the sum $x_1 = x + h$, in the transition to the 'differential' method we may view h 'as an *increment* (positive or negative) of x. This we have a right to do, since $x_1 - x = \Delta x$ and this same Δx can serve, after our way, as simply the symbol or sign of the differences of the x's, that is of $x_1 - x$, and also equally well as the quantity of the difference $x_1 - x$, as indeterminate as $x_1 - x$ and changed with their changing.

'Thus $x_1 - x = \Delta x$ or = the indeterminate quantity h. From this it follows that $x_1 = x + h$ and $f(x_1)$ or y_1 is transformed into $f(x + h)$.' (Yanovskaya, 1968 p.522)

In this way it would be unfair to represent the viewpoint of Marx as requiring the rejection of all other methods employed in differential calculus. If these methods are successful Marx sets himself the task of clarifying the secret of their success. And after this is shown to him, that is, after the examined method has demonstrated its validity and the conditions for its use are fulfilled, Marx considers a transition to this method not only fully justified but even appropriate.

Following his 1881 manuscript containing the fundamental results of his thoughts on the essence of differential calculus, Marx chose to send Engels a third work, concerned with the history of the method of differential calculus. At first, he wanted to depict this history with concrete examples of the various methods of showing the theorems on the derivation of the derivative, but then he *relinquished* this resolve and passed on to the general characteristics of the *fundamental* periods in the history of the methods of differential calculus.

This third work was not fully put into shape by Marx. There remain only the indications that he had decided to write about it and sketches of the manuscript, from which we know how Marx constructed and undertook the plan of his historical essay on this theme. This rough copy is published in full in the first part of this book (see pp.73-106). All of Marx's indications that there should be introduced into the text this or that page from other manuscripts are here followed in full. The manuscript gives us the possibility to explicate Marx's viewpoint on the history of the fundamental methods of differential calculus.

1) the 'mystical differential calculus' of Newton and Leibnitz,
2) the 'rational differential calculus' of Euler and d'Alembert,
3) the 'pure algebraic calculus' of Lagrange.

The characteristic features of the methods of Newton and Leibnitz revealed, according to Marx, the fact that their creators did not see the 'algebraic' kernel of differential calculus: they began immediately with their operational formulae, the origins and the meaning of which remained therefore misunderstood and even mysterious, so that the calculus stood out as 'a characteristic manner of calculation different from the usual algebra' (p.84), as a discovery, a completely special discipline of mathematics as 'different from the usual algebra as Heaven is wide' (p.113).

To the question, 'By what means . . . was the starting point chosen for the differential symbols as operational formulae' Marx answers, 'either through covertly or through overtly metaphysical assumptions, which themselves lead once more to metaphysical, unmathematical consequences, and so it is at that point that the violent suppression is made certain, the derivation is made to start its way, and indeed quantities made to proceed from themselves.' (p.64)

Elsewhere Marx writes concerning the methods of Newton and Leibnitz: '$x_1 = x + \Delta x$ from the beginning changes into $x_1 = x + dx$ . . . where dx is assumed by a metaphysical *explanation*. First, it exists, then it is explained.' 'From the arbitrary assumption the consequence follows that . . . terms . . . must be *juggled away*, in order to obtain the correct result.' (p.91)

In other words, so long as the meaning of *introduction* into mathematics of the differential symbols remains unexplained — more than that, generally false, since the differentials dx, dy are identified simply with the increments Δx, Δy — then the means of their *removal* appear unjustified, obtained by a 'forcible', 'juggling' suppression. We have to devise certain metaphysical, actually infinitely small quantities, which are to be treated *simultaneously* both as the usual different-from-zero (nowadays called 'Archimedean') quantities and as quantities which 'vanish' (transmute into zero) in comparison with the finite or infinitely small quantities of a lower order (that is, as 'non-Archimedean' quantities); or, simply put, as both zero and non-zero at the same time. 'Therefore nothing more remains,' writes Marx in this connection, 'than to imagine the increments h of the variable to be infinitely small increments and to give them as such *independent existence*, in the symbols $\dot{x}, \dot{y}$ etc. or dx, dy [etc] for example. But infinitely small quantities are quantities, just like those which are infinitely large (the word infinitely [small] only means in fact indefinitely small); the dy, dx . . . therefore also take part in the calculation just like ordinary algebraic quantities, and in the equation $(y+k)-y$ or $k = 2xdx + dxdx$ the $dxdx$ has the same right to existence as $2xdx$ does.' . . 'the reasoning is therefore most peculiar by which it is forcibly suppressed'. (p.83)

The presence of these actually infinitely small, that is, formally contradictory, items which are not introduced by means of operations of mathematically grounded consistency but are hypothesised on the basis of metaphysical 'explanations' and are removed by means of 'tricks' gives the calculus of Newton and Leibnitz, according to Marx, a 'mystical' quality, despite the many advantages they bring to it, thanks to which it begins immediately with operating formulae.

At the same time Marx rated very highly the *historical* significance of the methods of Newton and Leibnitz. 'Therefore,' he writes, 'mathematicians really believed in the mysterious character of the newly-discovered means of calculation which led to the correct (and, particularly in the geometric application, surprising) result by means of a positively false mathematical procedure. In this manner they became themselves mystified, rated the new discovery all the more highly, enraged all the more greatly the crowd of old orthodox mathematicians, and elicited the shrieks of hostility which echoed even in the world of non-specialists and which were necessary for the blazing of this new path.' (p.94)

The next stage in the development of the methods of differential calculus, according to Marx, was the 'rational differential calculus' of d'Alembert and Euler. The mathematically incorrect methods of Newton and Leibnitz are here corrected, but the starting point remains the same. 'D'Alembert starts directly from the *point de départ* of Newton and Leibnitz, $x_1 = x + dx$. But he immediately makes the fundamental correction: $x_1 = x + \Delta x$, that is x and an *undefined*, but prima facie *finite increment** which he calls h. The transformation of this h or Δx into dx . . . is the final result of the development, or

* By 'finite increment' the literature which Marx consulted understood a *non-zero* finite increment — *S.A. Yanovskaya*

at the least just before the gate swings shut, while in the mystics and the initiators of the calculus as its starting point.' (p.94) And Marx emphasised that with this the removal of the differential symbols from the final result proceeds then 'by means of correct mathematical operation. They are thus now discarded without sleight of hand.' (p.96)

Marx therefore rated highly the historical significance of d'Alembert's method. 'D'Alembert stripped the mystical veil from the differential calculus, and took an enormous step forward,' he writes (p.97).

However, so long as d'Alembert's starting point remains the representation of the variable x as the sum $x +$ an existing element, independent of the variable x, the increment Δx — then d'Alembert has not yet discovered the true dialectic process of differentiation. And Marx makes the critical observation regarding d'Alembert: 'D'Alembert begins with $(x + dx)$ but corrects the expression to $(x+\Delta x)$, alias $(x + h)$; a development now becomes necessary in which Δx or h is transformed into dx, but all of that development really proceeds.' (p.128)

As is well known, in order to obtain the result $\frac{dy}{dx}$ from the ratio of finite differences $\frac{\Delta y}{\Delta x}$, d'Alembert resorted to the 'limit process'. In the textbooks which Marx utilised, this passage to the limit foreshadowed the expansion of the expression $f(x + h)$ into all the powers of h, in which revealed in the coefficient of h raised to the first power was the 'already contained' derivative $f'(x)$.

The problem therefore became that of 'liberating' the derivative from the factor h and the other terms in the series. This was done naturally, so to speak, by simply defining the derivative as the coefficient of h raised to the first power in the expansion of $f(x + h)$ into a series of powers of h.

Indeed, 'in the first method *1*), as well as the rational one *2*), the real coefficient sought is fabricated ready-made by means of the binomial theorem; it is found at once in the second term of the series expansion, the term which therefore is necessarily combined with h^1. All the rest of the differential process then, whether in 1) or in 2), is a luxury. We therefore throw the needless ballast overboard.' (p.98)

The same thing was done by Lagrange, the founder of the next stage in the development of the differential calculus: 'pure algebraic' calculus, in Marx's periodisation.

At first Marx liked very much Lagrange's method, 'a theory of the derived function which gave a new foundation to the differential calculus'. Taylor's theorem, with which was usually obtained the expansion of $f(x + h)$ into a series of powers of h, and which historically arose as the crowning construction of the entire differential calculus, with this method was turned into the starting point of differential calculus, connecting it immediately with the mathematics preceding calculus (yet not employing its specific symbols). Marx noted with respect to this, 'the real and therefore the simplest interconnection of the new with the old is discovered as soon as the new gains its final form, and one may say, the differential calculus gained this relation through the theorems of Taylor and MacLaurin.* Therefore the thought first occurred to Lagrange to return the differential calculus to a firm algebraic foundation.' (p.113)

Marx found at once, however, that Lagrange did not make use of this insight. As is well known, Lagrange tried to show that 'generally speaking' — that is, with the exception of 'several special cases' in which differential calculus is 'inapplicable' — the expression $f(x + h)$ is expandable into the series

$$f(x) + ph + qh^2 + rh^3 + \dots,$$

where $p, q, r, \dots$ the coefficients for the powers of h, are new functions of x, independent of h, and 'derivable' from $f(x)$.

But Lagrange's proof of this theorem — in fact without much precise mathematic meaning — did not arise naturally. 'This leap from *ordinary algebra*, and besides *by means of ordinary* functions representing movement and change in general is as a *fait accompli*, it is not proved and is prima facie in *contradiction to all the laws* of conventional algebra . . . ' (p.177), writes Marx about this proof of Lagrange's.

And Marx concludes with respect to the 'initial equation' of Lagrange, that not only is it not proved, but also that 'the derivation of this equation from algebra therefore appears to rest on a deception' (p.117).

In the concluding part of the manuscript the method of Lagrange

* MacLaurin's Theorem can be regarded — as it was by Marx (pp.111, 112) — as a special case of Taylor's Theorem. — *Ed.*

appears as the completion of the method initiated by Newton and Leibnitz and corrected by d'Alembert; as the 'algebraicisation' based on Taylor by means of the method of formulae. 'In just such a manner Fichte followed Kant, Schelling Fichte, Hegel Schelling, and neither Fichte nor Schelling nor Hegel investigated the general foundations of Kant, of idealism in general: for otherwise they would not have been able to develop it further.' (p.119)

We can see that in a historical sketch Marx gives us a graphic example of what in his opinion should be the application of the method of dialectical materialism in such a science as the history of mathematics.

Completion of the present edition of *Mathematical Manuscripts* of Karl Marx required a great deal of preparation. The text of the manuscripts was translated in full; they were arranged chronologically; excerpts and summaries were separated from Marx's own statements; on the basis of analysis of their mathematical content the manuscripts were collected into units which can be read as a whole (in fact, many of the manuscripts do not make up notebooks, but are rather of separate sheets of paper in no sort of order). In the vast majority of cases it is known from which sources Marx drew his excerpts, or which he summarised. By comparison with the original works all of Marx's own comments have been identified in the summaries; all of Marx's independent work and notes have been translated into Russian.

The task of separating the personal opinions of Marx from his summaries and excerpts involved a series of difficulties. Marx wrote his summaries for his own benefit, in order to have at hand the material he needed. As always, he made use of a large collection of the most varied sources, but if he did not consider the account worth special attention, if it was, for example, a contemporary textbook compiled and widely distributed in England, then Marx very frequently did not accompany his excerpts with an indication of from where they were drawn. The task is complicated still further by the fact that the majority of the books which Marx utilised are now bibliographical rarities. In the final analysis all this work could only be completed at first hand in England, where, in order to resolve this problem, were studied and investigated in detail the stocks of the extant literature in these libraries: the British Museum, London and Cambridge universities, University College London, Trinity and St. James's Colleges in Cambridge, the Royal Society in London, and finally the private libraries of the eminent 19th century Englishmen de Morgan and Graves. Inquiries were made in other libraries as well, such as that of St. Catherine's College. For those manuscripts which by nature were prepared from German sources, the German historian of mathematics Wussing, at the request of the Institute, investigated the bibliographical resources of the German Democratic Republic.

Photocopies of several missing pages of the manuscripts were kindly provided by the Institute of Social History in Amsterdam, where the originals of the mathematical manuscripts of K. Marx are preserved.

Since the manuscripts are of the nature of rough drafts, one encounters omissions and even errors in the copied excerpts. The corresponding insertions or corrections are enclosed in square brackets. As a result the square brackets of Marx himself are indicated with double square brackets. Words which Marx abbreviated are written out in full, but the text is basically unchanged. In places obsolete orthography is even preserved.

The primary language of the manuscripts is German. If a reference in the manuscripts is in French or English, Marx sometimes writes his comments in French or English. In such cases Marx's text turns out to be so mixed that it becomes hard to say in what particular language the manuscript is written.

The dating of the manuscripts also entailed great difficulties. A detailed description of these difficulties is presented in the catalogue of manuscripts. This last lists the archival number of the manuscript, its assigned title, and the characteristics of either its sources or its content. Where the title or subtitle is Marx's own it is written in quotation marks in the original language and in Russian translation. In the first part of the book the titles not originating with Marx are marked with an asterisk.

The inventory of the manuscripts is given in the sequence of the arrangement of the archival sheets. Marx's own enumeration, by number or letters, is given in the inventory together with the indication of the archival sheets. An indication of the archival sheets on which they are found accompanies the published texts. All the manuscripts stem from fond 1, , opuscule 1.

The language of Marx's mathematical manuscripts in many cases departs from our usual contemporary language, and in order to understand his thought it is necessary to refer to the sources he used, to make clear the meaning of his terms. In order not to interrupt Marx's text, we place such explanations in the notes at the end of the book. Then, where more detailed information about the subject-matter of the sources consulted by Marx is found necessary, it is given in the Appendix. All such notes and references are of a purely informational character.

In Marx's texts are a great number of underlinings, by means of which he emphasised the points of particular importance to him. All these underlinings are indicated by means of italics.

The book was prepared by S.A. Yanovskaya, professor of the M.V. Lomonosov Moscow Government University, to whom also are due the Preface, the Inventory of mathematical manuscripts (compiled with the assistance of A.Z. Rybkin), the Appendices and the Notes. Professor K.A. Rybnikov took part in the editing of the book, performing among other tasks the greater part of the work of researching the sources used by K. Marx in his work on the 'Mathematical Manuscripts'. In the preparation of the present edition the comments and advice of Academicians A.N. Kolmogorov and I.G. Petrovskii were carefully considered.

A.Z. Rybkin, chief editor for the physical-mathematical section of Nauka Press, and O.K. Senekina, of the Institute for Marxism-Leninism of the Central Committee of the Communist Party of the Soviet Union, directed all the work of editing the book, preparing it for publication and proof-reading it. The book includes an index of references quoted and consulted, as well as an index of names. References in Marx's text are denoted in the indices by means of italics.

Two Manuscripts on Differential Calculus

I

'ON THE CONCEPT OF THE DERIVED FUNCTION'[1]

I

Let the independent variable x increase to x_1; then the dependent variable y increases to y_1.[2]

Here in I) we consider the simplest possible case, where x appears only to the first power.

1) $y = ax$; when x increases to x_1,

$$y_1 = ax_1 \text{ and } y_1 - y = a(x_1 - x) .$$

Now allow the *differential operation* to occur, that is, we let x_1 take on the value of x. Then

$$x_1 = x ; \qquad x_1 - x = 0 ,$$

thus

$$a(x_1 - x) = a . 0 = 0 .$$

Furthermore, since y only becomes y_1 because x increases to x_1, we have at the same time

$$y_1 = y ; \qquad y_1 - y = 0 .$$

Thus

$$y_1 - y = a(x_1 - x)$$

changes to $0 = 0$.

First making the differentiation and then removing it therefore leads literally to *nothing*. The whole difficulty in understanding the differential operation (as in the *negation of the negation* generally) lies precisely in seeing *how* it differs from such a simple procedure and therefore leads to real results.

If we divide both $a(x_1 - x)$ and the left side of the corresponding equation by the factor $x_1 - x$, we then obtain

$$\frac{y_1 - y}{x_1 - x} = a .$$

Since y is the *dependent variable*, it cannot carry out any independent motion at all, y_1 therefore cannot equal y and $y_1 - y = 0$ without x_1 first having become equal to x.

On the other hand we have seen that x_1 cannot become equal to x in the function $a(x_1 - x)$ without making the latter $= 0$. The factor $x_1 - x$ was thus *necessarily* a *finite difference*[3] when both sides of the equation were divided by it. At the moment of the construction of the ratio

$$\frac{y_1 - y}{x_1 - x}$$

$x_1 - x$ is therefore always a finite difference. It follows that

$$\frac{y_1 - y}{x_1 - x}$$

is a *ratio of finite differences*, and correspondingly

$$\frac{y_1 - y}{x_1 - x} = \frac{\Delta y}{\Delta x}$$

Therefore

$$\frac{y_1 - y}{x_1 - x} \text{ or}^4 \ \frac{\Delta y}{\Delta x} = a ,$$

where the constant a represents the *limit value* (*Grenzwert*) of the ratio of the finite differences of the variables.[5]

Since a is a constant, no change may take place in it; hence none can occur on the *right-hand side* of the equation, which has been reduced to a. Under such circumstances the *differential process* takes place on the left-hand side

$$\frac{y_1 - y}{x_1 - x} \text{ or } \frac{\Delta y}{\Delta x} ,$$

and this is characteristic of such simple functions as ax.

If in the denominator of this ratio x_1 decreases so that it approaches x, the limit of its decrease is reached as soon as it becomes x. Here the difference becomes $x_1 - x_1 = x - x = 0$ and therefore also $y_1 - y = y - y = 0$. In this manner we obtain

$$\frac{0}{0} = a \ .$$

Since in the expression $\frac{0}{0}$ every trace of its origin and its meaning has disappeared, we replace it with $\frac{dy}{dx}$, where the finite differences $x_1 - x$ or Δx and $y_1 - y$ or Δy appear symbolised as *cancelled* or *vanished* differences, or $\frac{\Delta y}{\Delta x}$ changes to $\frac{dy}{dx}$.

Thus

$$\frac{dy}{dx} = a \ .$$

The closely-held belief of some rationalising mathematicians that dy and dx are quantitatively actually only infinitely small, only approaching $\frac{0}{0}$, is a chimera, which will be shown even more palpably under II).

As for the characteristic mentioned above of the case in question, the limit value (*Grenzwert*) of the finite differences is therefore also at the same time the limit value of the differentials.

2) A second example of the same case is

$$\begin{array}{l} y = x \\ y_1 = x_1 \end{array} \ ; \qquad y_1 - y = x_1 - x \ ;$$

$$\frac{y_1 - y}{x_1 - x} \text{ or } \frac{\Delta y}{\Delta x} = 1 \ ; \qquad \frac{0}{0} \text{ or } \frac{dy}{dx} = 1 \ .$$

II

When in $y = f(x)$, the function [of] x appears on the right-hand side of the equation in its *developed algebraic expression*,[6] we call this expression the *original function of x*, its first modification obtained by means of differentiation the *preliminary 'derived' function of x* and its final form obtained by means of the *process of differentiation* the *'derived' function of x*.[7]

1) $y = ax^3 + bx^2 + cx - e$.

If x increases to x_1, then

$$y_1 = ax_1^3 + bx_1^2 + cx_1 - e\,,$$

$$\begin{aligned} y_1 - y &= a(x_1^3 - x^3) + b(x_1^2 - x^2) + c(x_1 - x) \\ &= a(x_1 - x)(x_1^2 + x_1x + x^2) \\ &\quad + b(x_1 - x)(x_1 + x) + c(x_1 - x)\,. \end{aligned}$$

Therefore

$$\frac{y_1 - y}{x_1 - x} \text{ or } \frac{\Delta y}{\Delta x} = a(x_1^2 + x_1x + x^2) + b(x_1 + x) + c\,.$$

and the *preliminary 'derivative'* [is]

$$a(x_1^2 + x_1x + x^2) + b(x_1 + x) + c$$

[and it] is here the *limit value* (*Grenzwert*) of the *ratios* of the finite differences; that is, however small these differences may become, the value of $\frac{\Delta y}{\Delta x}$ is given by that 'derivative'. But this is not the same case as that under I) with the limit value of the ratios of the differentials.*

* In a draft of this work (4146, Pl.4), the following appears: 'On the other hand, the process of differentiation (*Differentialprozess*) now takes place in the preliminary "derived" function of *x* (on the right-hand side), while any movement of the same process on [the] left-hand side is necessarily prohibited.' — *Ed.*

When the variable x_1 is decreased in the function

$$a(x_1^2 + x_1 x + x^2) + b(x_1 + x) + c$$

until it has reached the limit of its decrease, that is, has become *the same as x*, [then] x_1^2 is changed to x^2, $x_1 x$ to x^2, and $x_1 + x$ to $2x$, and we obtain the *'derived' function* of x:

$$3ax^2 + 2bx + c\ .$$

It is here shown in a striking manner:

First: in order to obtain the 'derivative', x_1 must be set $= x$; therefore in the *strict mathematical sense* $x_1 - x = 0$, with no subterfuge about merely approaching infinitely [closely].

Second: Although we set $x_1 = x$ and therefore $x_1 - x = 0$, nonetheless nothing symbolic appears in the 'derivative'.* The quantity x_1, although originally obtained from the variation of x, does not disappear; it is only reduced to its minimum limit value $= x$. It remains in the original function of x as a newly introduced element which, by means of its combinations partly with itself and partly with the x of the original function, finally produces the 'derivative', that is, the *preliminary derivative reduced* to its *absolute minimum quantity*.

The reduction of x_1 to x within the first (preliminary) 'derived' function changes the left-hand side [from] $\frac{\Delta y}{\Delta x}$ to $\frac{0}{0}$ or $\frac{dy}{dx}$, thus:

$$\frac{0}{0} \text{ or } \frac{dy}{dx} = 3ax^2 + 2bx + c\ ,$$

* The draft contains the following statement: 'Finding "the derivative" from the original function of x proceeds in such a manner, that we first take a *finite differentiation* (*endliche Differentiation*); this provides a preliminary "derivative" which is the *limit value* (*Grenzwert*) of $\frac{\Delta y}{\Delta x}$. The process of differentiation (*Differentialprozess*) to which we then proceed, *reduces* this *limit value* to its absolute minimum quantity (*Minimalgrösse*). The quantity x_1 introduced in the first differentiation does not disappear . . .' — *Ed.*

so that the *derivative* appears as the *limit value* of the ratio of the differentials.

The transcendental or symbolic mistake which appears only on the left-hand side has perhaps already lost its terror since it now appears only as the expression of a process which has established its real content on the right-hand side of the equation.

In the 'derivative'

$$3ax^2 + 2bx + c$$

the variable x exists in a completely different condition than in the original function of x (namely, in $ax^3 + bx^2 + cx - e$). It [this derivative] can therefore itself be treated as an original function in turn, and can become the mother of another 'derivative' by the repeated process of differentiation. This can be repeated as long as the variable x has not been finally removed from one of the 'derivatives'; it therefore continues endlessly in functions of x which can only be represented by infinite series, which [is] all too often the case.

The symbols $\frac{d^2y}{dx^2}$, $\frac{d^3y}{dx^3}$, etc., only display the genealogical register of the 'derivatives' with respect to the original given function of x. They are mysterious only so long as one treats them as the *starting point* of the exercise, instead of as merely *the expressions of the successively derived functions of* x. For it indeed appears miraculous that a ratio of vanished quantities should pass through a new, higher degree of disappearance, while there is nothing wonderful in the fact that $3x^2$, for example, can pass through the process of differentiation as well as its mother x^3. One could just as well begin with $3x^2$ as with the original function of x.

But *nota bene*. The starting point of the *process of differentiation* actually is $\frac{\Delta y}{\Delta x}$ only in equations as [above] under I), where x appears only to the first power. Then, however, as was shown under I), the result [is]:

$$\frac{\Delta y}{\Delta x} = a = \frac{dy}{dx}.$$

Here therefore as a matter of fact *no new limit value* is found from the process of differentiation which $\frac{\Delta y}{\Delta x}$ passes through; [a result] which remains possible only so long as the preliminary 'derivative' includes the variable x, so long, therefore, as $\frac{dy}{dx}$ remains the symbol of a real process.*

Of course, it is in no sense an obstacle, that in the differential calculus the symbols $\frac{dy}{dx}$, $\frac{d^2y}{dx^2}$, etc., and their combinations also appear on the right-hand side of the equation. For one knows as well that such purely symbolic equations only indicate the *operations* which are then to be applied to the real functions of variables.

2) $y = ax^m$.

As x becomes x_1, then $y_1 = ax_1^m$ and

$$y_1 - y = a(x_1^m - x^m)$$
$$= a(x_1 - x)\,(x_1^{m-1} + x_1^{m-2}x + x_1^{m-3}x^2 + \text{ etc.}$$
$$\text{up to the term } x_1^{m-m}\,x^{m-1}).$$

Therefore

$$\frac{y_1 - y}{x_1 - x} \text{ or } \frac{\Delta y}{\Delta x} = a(x_1^{m-1} + x_1^{m-2}x + x_1^{m-3}x^2 + \ldots + x_1^{m-m}x^{m-1}).$$

We now apply the process of differentiation to this *'preliminary derivative'*, so that

* The draft (Pl.7) includes this sentence: 'This can only come about, where the preliminary "derived" function includes the variable x, through whose motion, therefore, another truly new value may be formed, so that $\frac{dy}{dx}$ is the symbol of a real process.' — *Ed.*

$$x_1 = x \text{ or } x_1 - x = 0$$

and

$$x_1^{m-1} \text{ is changed into } x^{m-1};$$

$$x_1^{m-2}x \text{ into } x^{m-2}x = x^{m-2+1} = x^{m-1};$$

$$x_1^{m-3}x^2 \text{ into } x^{m-3}x^2 = x^{m-3+2} = x^{m-1},$$

and finally,

$$x_1^{m-m}\, x^{m-1} \text{ into } x^{m-m}\, x^{m-1} = x^{0+m-1} = x^{m-1}.$$

We thus obtain the function x^{m-1} m times, and the 'derivative' is therefore max^{m-1}.

Due to the equivalence of $x_1 = x$ within the 'preliminary derivative',* on the left-hand side $\frac{\Delta y}{\Delta x}$ is changed to $\frac{0}{0}$ or $\frac{dy}{dx}$; therefore

$$\frac{dy}{dx} = max^{m-1}.$$

All of the operations of the differential calculus could be treated in this manner, which would however be a damned useless mass of details. Nonetheless here is another example; since up to now the difference $x_1 - x$ appeared *only once* in the function of x and therefore disappeared from the right-hand side by means of the formation of

$$\frac{y_1 - y}{x_1 - x} \text{ or } \frac{\Delta y}{\Delta x}.$$

This [is] not the case in the following:

3) $y = a^x$;
Let x become x_1. Then

$$y_1 = a^{x_1}$$

*On the right-hand side, that is. — *Ed.*

Therefore

$$y_1 - y = a^{x_1} - a^x = a^x(a^{x_1-x} - 1)\ .$$

[But]

$$a^{x_1-x} = \{1 + (a-1)\}^{x_1-x}\ ,$$

and

$$\{1 + (a-1)\}^{x_1-x} =$$

$$1 + (x_1 - x)(a-1) + \frac{(x_1-x)(x_1-x-1)}{1.2}(a-1)^2 + \text{etc.}$$ [8]

Therefore

$$y_1 - y = a^x(a^{x_1-x} - 1)$$

$$= a^x\left\{(x_1-x)(a-1) + \frac{(x_1-x)(x_1-x-1)}{1.2}(a-1)^2 + \frac{(x_1-x)(x_1-x-1)(x_1-x-2)}{1.2.3}(a-1)^3 + \text{etc.}\right\}.$$

$\therefore$ [9] $$\frac{y_1-y}{x_1-x} \text{ or } \frac{\Delta y}{\Delta x} =$$

$$a^x\left\{(a-1) + \frac{x_1-x-1}{1.2}(a-1)^2 + \frac{(x_1-x-1)(x_1-x-2)}{1.2.3}(a-1)^3 + \text{etc.}\right\}.$$

Now as $x_1 = x$ and thus $x_1 - x = 0$, we obtain for the 'derivative':

$$a^x\left\{(a-1) - \frac{1}{2}(a-1)^2 + \frac{1}{3}(a-1)^3 - \text{etc.}\right\}.$$

Thus

$$\frac{dy}{dx} = a^x \left\{(a-1) - \frac{1}{2}(a-1)^2 + \frac{1}{3}(a-1)^3 - \text{ etc.}\right\}$$

If we designate the sum of the constants in parentheses A, then

$$\frac{dy}{dx} = Aa^x \ ;$$

but this A = the Napierian logarithm of the number* a, so that:

$\frac{dy}{dx}$, or, when we replace y by its value: $\frac{da^x}{dx} = \log a \,.\, a^x$,

and

$$da^x = \log a \,.\, a^x dx \ .$$

Supplementary[10]
We have considered

1) cases in which the factor $(x_1 - x)$ [occurs] only once in [the expression which leads to] the '*preliminary derivative*' — i.e. [in] the equation of finite differences[11] — so that by means of the division of both sides by $x_1 - x$ in the formation of

$$\frac{y_1 - y}{x_1 - x} \text{ or } \frac{\Delta y}{\Delta x}$$

this same factor is therefore eliminated from the function of x.

2) (in the example $d(a^x)$) cases in which factors of $(x_1 - x)$ remain after the formation of $\frac{\Delta y}{\Delta x}$.[12]

3) Yet to be considered is the case where the factor $x_1 - x$ is *not directly* obtained from the first difference equation ([which leads to] the 'preliminary derivative').

* Original: 'root'. — *Trans.*

$$y = \sqrt{a^2 + x^2}\,,$$
$$y_1 = \sqrt{a^2 + x_1^2}\,,$$
$$y_1 - y = \sqrt{a^2 + x_1^2} - \sqrt{a^2 + x^2}\,;$$

we divide the function of x, the left-hand side as well, therefore, by $x_1 - x$. Then

$$\frac{y_1 - y}{x_1 - x}\left(\text{or } \frac{\Delta y}{\Delta x}\right) = \frac{\sqrt{a^2 + x_1^2} - \sqrt{a^2 + x^2}}{x_1 - x}\,.$$

In order to rationalise the numerator, [both] numerator and denominator are multiplied by $\sqrt{a^2 + x_1^2} + \sqrt{a^2 + x^2}$, and we obtain:

$$\frac{\Delta y}{\Delta x} = \frac{a^2 + x_1^2 - (a^2 + x^2)}{(x_1 - x)(\sqrt{a^2 + x_1^2} + \sqrt{a^2 + x^2})}$$

$$= \frac{x_1^2 - x^2}{(x_1 - x)(\sqrt{a^2 + x_1^2} + \sqrt{a^2 + x^2})}\,.$$

But

$$\frac{x_1^2 - x^2}{(x_1 - x)(\sqrt{a^2 + x_1^2} + \sqrt{a^2 + x^2})}$$

$$= \frac{(x_1 - x)(x_1 + x)}{(x_1 - x)(\sqrt{a^2 + x_1^2} + \sqrt{a^2 + x^2})}\,.$$

So that:

$$\frac{\Delta y}{\Delta x} = \frac{x_1 + x}{\sqrt{a^2 + x_1^2} + \sqrt{a^2 + x^2}}.$$

Now when x_1 becomes $= x$, or $x_1 - x = 0$, then

$$\frac{dy}{dx} = \frac{2x}{2\sqrt{a^2 + x^2}} = \frac{x}{\sqrt{a^2 + x^2}}.$$

So that

$$dy \text{ or } d\sqrt{a^2 + x^2} = \frac{xdx}{\sqrt{a^2 + x^2}}.$$

ON THE DIFFERENTIAL[13]

I

1) Let $f(x)$ or $y = uz$ be a function to be differentiated; u and z are both functions dependent on the independent variable x. They are independent variables with respect to the function y, which depends on them, and thus on x.

$$y_1 = u_1 z_1,$$

$$y_1 - y = u_1 z_1 - uz = z_1(u_1 - u) + u(z_1 - z),$$

$$\frac{y_1 - y}{x_1 - x} \text{ or } \frac{\Delta y}{\Delta x} = z_1 \frac{u_1 - u}{x_1 - x} + u \frac{z_1 - z}{x_1 - x} = \frac{z_1 \Delta u}{\Delta x} + \frac{u \Delta z}{\Delta x}.\star$$

Now on the right-hand side let $x_1 = x$, so that $x_1 - x = 0$, likewise $u_1 - u = 0$, $z_1 - z = 0$; so that the factor z_1 in $z_1 \frac{u_1 - u}{x_1 - x}$ also goes to z; finally on the left-hand side $y_1 - y = 0$. Therefore:

A) $\dfrac{dy}{dx} = z \dfrac{du}{dx} + u \dfrac{dz}{dx}$.

Which equation, when all its terms are multiplied by the common denominator dx, becomes

B) dy or $d(uz) = z\,du + u\,dz$.[14]

2) Consider for the time being the first equation A):

$$\frac{dy}{dx} = z \frac{du}{dx} + u \frac{dz}{dx}.$$

⋆ The last part of the equation was apparently added by Engels — *Ed.*

In equations with only one variable dependent on x, the final result has always been

$$\frac{dy}{dx} = f'(x) \, ,$$

and $f'(x)$, the first derived function* of $f(x)$, has been free[15] of all symbolic expressions, for example, mx^{m-1} when x^m is the original function of the independent variable x. As a direct result of the process of differentiation which $f(x)$ had to pass through in order to be transformed into $f'(x)$, its shadow image (*Doppelgänger*) $\frac{0}{0}$ or $\frac{dy}{dx}$ appeared as the symbolic equivalent on the left-hand side opposite $f'(x)$, the real differential coefficient.[16] Alternately $\frac{0}{0}$ or $\frac{dy}{dx}$ found its real equivalent in $f'(x)$.

In equation A) by contrast, $f'(x)$, the first derivative of uz, itself includes symbolic differential coefficients, which are therefore present on both sides while on neither is there a real value. Since, however, uz has been handled in the same manner as the earlier functions of x with only one independent variable, this contrast is obviously a result of the peculiar character of the beginning function itself, namely uz. A more complete treatment of this is found under 3).

For the moment, it remains to be seen whether there are any twists in the *derivation* of equation A).

On the right-hand side

$$\frac{u_1 - u}{x_1 - x} \quad \text{or} \quad \frac{\Delta u}{\Delta x} \quad \text{and} \quad \frac{z_1 - z}{x_1 - x} \quad \text{or} \quad \frac{\Delta z}{\Delta x}$$

become $\frac{0}{0}$, $\frac{0}{0}$, because x_1 has become $= x$, so that $x_1 - x = 0$.

In place of $\frac{0}{0}$, $\frac{0}{0}$ we put $\frac{du}{dx}$, $\frac{dz}{dx}$ without further ado. Was that permissible, since these $\frac{0}{0}$ figure here as the *multipliers of the variables* u and z respectively, while in cases with one

* Synonymous with 'derivative' — *Trans.*

independent variable the single symbolic differential coefficient — $\frac{0}{0}$ or $\frac{dy}{dx}$ — has no multiplier other than the constant, 1?

If we place the primitive problematic form of $\frac{du}{dx}$, $\frac{dz}{dx}$ on the right-hand side it becomes: $z\frac{0}{0} + u\frac{0}{0}$. If we then multiply z and u by the numerators of the $\frac{0}{0}$ accompanying them, we obtain: $\frac{0}{0} + \frac{0}{0}$; and since the variables z and u themselves become $= 0$,[17] as are their derivatives as well, so that [we obtain] finally:

$$\frac{0}{0} = 0 \quad \text{and not} \quad z\frac{du}{dx} + u\frac{dz}{dx}.$$

This procedure, however, is mathematically false.

Let us take, for example

$$\frac{u_1 - u}{x_1 - x} = \frac{\Delta u}{\Delta x};$$

one does not first obtain the numerator $= 0$ because one has begun with it and set $u_1 - u = 0$, but rather the numerator only becomes 0 or $u_1 - u = 0$ because the denominator, the difference of the independent variable quantities x, that is $x_1 - x$, has become $= 0$.

Therefore what arises opposite the variables u and z is not 0 but $\left(\frac{0}{0}\right)$, *whose numerator in this form remains inseparable from its denominator*. Consequently as a multiplier $\frac{0}{0}$ then could nullify its coefficients only when and so far as

$$\frac{0}{0} = 0.$$

Even in the usual algebra it would be false, in the case where a product $P.\frac{m}{n}$ takes the form $P.\frac{0}{0}$, to conclude immediately that it *must be* $= 0$, although it *may be* set always $= 0$ here, since we can begin[18] the nullification arbitrarily with numerator or denominator.

For example, $P.\frac{x^2 - a^2}{x - a}$. Let [because $x = a$] x^2 be set $= a^2$, so that $x^2 - a^2 = 0$; we then obtain: $P.\frac{0}{0} = \frac{0}{0}$, and the last [term] may be set $= 0$, since $\frac{0}{0}$ can just as readily be 0 as any other number.

By contrast, let us reduce $x^2 - a^2$ to its factors, so that we obtain

$$P.\frac{x - a}{x - a}.(x + a) = P(x + a),\text{ and since } x = a,^{19} = 2Pa .$$

Successive differentiation — for example, of x^3, where $\frac{0}{0}$ first becomes $= 0$ only in the fourth derivative, since in the third the variable x has run out and is replaced by a constant — proves that $\frac{0}{0}$ becomes $= 0$ only under completely defined conditions.

In our case, however, where the origin of $\frac{0}{0}, \frac{0}{0}$ is known to be the differential expression of $\frac{\Delta z}{\Delta x}, \frac{\Delta u}{\Delta x}$ respectively, the two deserve, as above, the 'uniform' (*die Uniform*) $\frac{dz}{dx}, \frac{du}{dx}$.

3) In the equations, such as $y = x^m$, $y = a^x$ etc., which have been treated previously, an *original function of* x stands opposite a y 'dependent' on it.

In $y = uz$, both sides contain 'dependent [variables]'. While here y depends directly on u and z, so in turn u and z [depend] as well on x. This specific character of the original function uz necessarily stamps on its 'derivatives' as well.

That u is a function of x, and z another function of x is represented by:

$$u = f(x)\,, \qquad u_1 - u = f(x_1) - f(x)\,,$$

and

$$z = \varphi(x)\,; \qquad z_1 - z = \varphi(x_1) - \varphi(x)\,.$$

But neither the beginning equation for $f(x)$ nor for $\varphi(x)$ leads to an original function of x, that is, a definite value* in x. Consequently u and z figure as mere names, as symbols of functions of x; therefore as well only the *general forms of this ratio of dependence* (*Abhängigkeitsverhältnis*) :

$$\frac{u_1 - u}{x_1 - x} = \frac{f(x_1) - f(x)}{x_1 - x}\,, \qquad \frac{z_1 - z}{x_1 - x} = \frac{\varphi(x_1) - \varphi(x)}{x_1 - x}$$

is generated immediately by the process of taking the derivative. The process has now reached the point where x_1 is set = x, so that $x_1 - x = 0$, and those general forms are transformed to

$$\frac{du}{dx} = \frac{df(x)}{dx}\,, \qquad \frac{dz}{dx} = \frac{d\varphi(x)}{dx}\,,$$

and the symbolic differential coefficients $\frac{du}{dx}$, $\frac{dz}{dx}$ become as such incorporated into the 'derivatives'.

In equations with only one dependent variable, $\frac{dy}{dx}$ has no other content at all than $\frac{du}{dx}$, $\frac{dz}{dx}$ have here. It is also merely the symbolic differential expression of

$$\frac{y_1 - y}{x_1 - x} = \frac{f(x_1) - f(x)}{x_1 - x}\,.$$ [20]

Although the nature of $\frac{du}{dx}$, $\frac{dz}{dx}$ — that is, of symbolic coefficients in general — is in no way altered when they appear *within the derivative itself*, and so on the right-hand side of the

---

* 'Definite expression' is meant — *Ed.*

differential equation as well, nonetheless their role and the character of the equation are thereby altered.

Let us represent the original function of uz, in combination, by $f(x)$, and their first 'derivative' by $f'(x)$,

$$\frac{dy}{dx} = z\,\frac{du}{dx} + u\,\frac{dz}{dx}$$

then becomes:

$$\frac{dy}{dx} = f'(x)\,.$$

We have obtained this very general form for equations with only one dependent variable. In both cases the beginning forms of $\frac{dy}{dx}$ arose from the process of taking the derivative (*Ableitungsprozesse*), which transforms $f(x)$ into $f'(x)$. So soon, therefore, as $f(x)$ becomes $f'(x)$, $\frac{dy}{dx}$ stands opposite the latter as its own symbolic expression, as its shadow image (*Doppelgänger*) or symbolic equivalent.

In both cases, therefore, $\frac{dy}{dx}$ plays *the same role*.

It is otherwise with $\frac{du}{dx}$, $\frac{dz}{dx}$. Together with the other elements of $f'(x)$, into which they are incorporated, in $\frac{dy}{dx}$ they meet with their symbolic expression or their symbolic equivalent, but they themselves do not stand opposite the $f'(x)$, $\varphi'(x)$ whose symbolic shadow images they would be in turn. They are brought into the world unilaterally, shadow figures lacking the body which cast them, symbolic differential coefficients without the real differential coefficients, that is, without the corresponding equivalent 'derivative'. Thus the symbolic differential coefficient becomes the *autonomous starting point* whose real equivalent is first to be found. The initiative is thus shifted from the right-hand pole, the algebraic, to the left-hand one, the symbolic. Thereby, however, the differential calculus also appears as a specific type of calculation which already operates independently on its own ground (*Boden*). For its starting points $\frac{du}{dx}$, $\frac{dz}{dx}$ belong only to it and are mathematical quantities characteristic of it. And this inversion of the method arose as a result of the algebraic differentiation of uz. The algebraic method therefore inverts itself into its exact opposite, the differential method.*

Now, what are the corresponding 'derivatives' of the symbolic differential coefficients $\frac{du}{dx}$, $\frac{dz}{dx}$? The beginning equation $y = uz$ provides no data for the resolution of this question. This last [question] may still be answered if one substitutes arbitrary original functions of x for u and z. For example,

$$u = x^4 \;; \qquad z = x^3 + ax^2 \;.$$

Thereby, however, the symbolic differential coefficients $\frac{du}{dx}$, $\frac{dz}{dx}$ are suddenly transformed into *operational symbols* (*Operationssymbole*), into symbols of the process which must be carried out with x^4 and $x^3 + ax^2$ in order to find their 'derivatives'. Originally having arisen as the symbolic expression of the 'derivative' and thus already finished, the symbolic differential coefficient now plays the role of the symbol of the

---

* The draft of the work 'On the Differential' (4148, Pl.16-17) contains this paragraph:

'$\frac{du}{dx}$, $\frac{dz}{dx}$ thrown over. Born within the derivative, they, together with the remaining elements of the same, meet in $\frac{dy}{dx}$ their own symbolic expression, therefore their symbolic equivalent. But they themselves exist without equivalent, real differential coefficients, that is without the derivative $f'(x)$, $\varphi'(x)$ whose symbolic expression they in turn had been. They are the completed differential symbols whose real values figure as shadows whose bodies are to be sought first. The problem has thus been turned around before one's eyes. The symbolic differential coefficients have become autonomous *starting points*, for whom the equivalent, the real differential coefficient or the corresponding derived function, is first to be sought. Thereby the initiative has been shifted from the right-hand pole to the left. Since this inversion of the method originated from the algebraic manipulation of the function uz, it has itself been demonstrated algebraically.' — *Ed.*

operation of differentiation which is yet to be completed.

At the same time the equation

$$\frac{dy}{dx} = z\frac{du}{dx} + u\frac{dz}{dx},$$

from the beginning purely symbolic, because lacking a side free of symbols, has been transformed into a general symbolic operational equation.

I remark further that* from the early part of the 18th century right down to the present day, the general task of the differential calculus has usually been formulated as follows: to find the real equivalent of the symbolic differential coefficient.

4)

A) $\frac{dy}{dx} = z\frac{du}{dx} + u\frac{dz}{dx}.$

This is obviously not the simplest expression of equation A), since all its terms have the denominator dx in common. Let this be struck out, and then:

B) $d(uz)$ or $dy = zdu + udz$.

Any trace in B) of its origin in A) has disappeared. It is therefore equally as valid when u and z depend on x as when they depend only reciprocally on one another, without any relation to x at all.[21] From the beginning it has been a symbolic equation and from the beginning could have served as a symbolic operational equation. In the present case it means, that when

$$y = zu \text{ etc.},$$

that is = a product of any arbitrary number of variables multiplied together, then dy = a sum of products, in each one of which one of the factors is treated as a variable while the other factors are treated as constants, etc.

For our purpose, namely the further investigation of the differential of y in general, form B) nonetheless will not do. We therefore set:

* The following is in the draft: 'save for a few exceptions'. — *Ed.*

$$u = x^4, \qquad z = x^3 + ax^2 .$$

so that

$$du = 4x^3dx , \qquad dz = (3x^2 + 2ax)dx ,$$

as was proved earlier for equations with only one dependent variable. These values of du , dz are brought into equation A), so that

A) $\dfrac{dy}{dx} = (x^3 + ax^2)\dfrac{4x^3dx}{dx} + x^4\dfrac{(3x^2 + 2ax)dx}{dx}$; and then

$$\frac{dy}{dx} = (x^3 + ax^2)4x^3 + x^4(3x^2 + 2ax) ;$$

therefore

$$dy = \{(x^3 + ax^2)4x^3 + x^4(3x^2 + 2ax)\}\, dx .$$

The expression in brackets is the first derivative of uz; since, however, $uz = f(x)$, its derivative is $= f'(x)$; we now substitute the latter in place of the algebraic function, and so:

$$dy = f'(x)dx .$$

We have already obtained the same result from an arbitrary equation with only one variable. For example:

$$y = x^m ,$$

$$\frac{dy}{dx} = mx^{m-1} = f'(x) ,$$

$$dy = f'(x)dx .$$

In general we have: if $y = f(x)$, whether this function of x is now an original function in x or contains a dependent variable, then always $dy = df(x)$ and $df(x) = f'(x)dx$, and so:

B) $dy = f'(x)dx$ is the most generally valid form of the differential of y. This would be demonstrable immediately also if the given $f(x)$ were $f(x, z)$, that is a function *of two mutually independent variables*. For our purposes, however, this would be superfluous.

II

1) The differential

$$dy = f'(x)dx$$

appears right away to be more suspicious than the differential coefficient

$$\frac{dy}{dx} = f'(x)$$

from which it is derived.

In $\frac{dy}{dx} = \frac{0}{0}$ the numerator and denominator are inseparably bound; in $dy = f'(x)dx$ they are apparently separated, so that one is forced to the conclusion that it is only a disguised expression for

$$0 = f'(x).0 \text{ or } 0 = 0 ,$$

whereupon 'nothing's to be done' ('*nix zu wolle*').

A French mathematician of the first third of the 19th century, who is clear in a completely different manner than the well-known [to you] 'elegant' Frenchman,[22] has drawn a connection between the differential method and Lagrange's algebraic method: — Boucharlat says:

If for example $\frac{dy}{dx} = 3x^2$, then '$\frac{dy}{dx}$ alias $\frac{0}{0}$, or rather its value $3x^2$, is the differential coefficient of the function y. Since $\frac{dy}{dx}$ is thus the symbol which represents the value $3x^2$, ***dx must always stay (stehn)**** ***under dy***, but ***in order to facilitate algebraic operation*** we treat $\frac{dy}{dx}$ as an ordinary fraction and $\frac{dy}{dx} = 3x^2$ as an ordinary equation. By removing the denominator from the equation one obtains the result

$$dy = 3x^2 dx ,$$

which expression is called the differential of y'.[23]

* The draft has: 'remain' (*stehn bleiben*) — *Ed.*

Thus, in order 'to facilitate algebraic operation', one introduces a demonstrably false formula which one baptises the 'differential'.

In fact the situation is not so nasty.

In $\frac{0}{0}$* the numerator is inseparable from the denominator, but why? Because both only express a ratio if they are not separated, something like (*dans l'espèce*) the ratio[24] reduced to its absolute minimum:

$$\frac{y_1 - y}{x_1 - x} = \frac{f(x_1) - f(x)}{x_1 - x},$$

where the numerator goes to 0 because the denominator has done so. Separated, both are 0; they lose their symbolic meaning, their reason.

As soon, however, as $x_1 - x = 0$ achieves in dx a form which is manifested without modification as the vanished difference in the independent variable x, so that dy as well is a vanished difference in the function of x or in the dependent [variable] y, then the separation of the denominator from the numerator becomes a completely permissible operation. Wherever dx stands now, such a change of position leaves the ratio of dy to dx undisturbed. $dy = f'(x)dx$ thus appears to us to be an alternative form of

$$\frac{dy}{dx} = f'(x)$$

and may always be substituted for the latter.[25]

2) The differential $dy = f'(x)dx$ arose from A) by means of a direct algebraic derivation (see I,4), while the algebraic derivation of equation A) had already shown that the differential symbol, somewhat like (*dans l'espèce*) the symbolic differential coefficient which originally emerged as a purely symbolic expression of the algebraically performed process of differentiation, necessarily inverts into an independent starting

* The draft has: 'In the form $\frac{0}{0}$' — *Ed.*

point, into a symbol of an operation yet to be performed, into an operational symbol, and thus the symbolic equations which have emerged along the algebraic route also invert into symbolic operational equations (*Operationsgleichungen*).

We are thus doubly correct in treating the differential $y = f'(x)dx$ as a symbolic operational equation. So we now know *a priori*, that if

$$y = f(x) \quad [\text{then}] \quad dy = df(x) ,$$

that if the operation of differentiation indicated by $df(x)$ is performed on $f(x)$, the result is $dy = f'(x)dx$, and that from this results finally

$$\frac{dy}{dx} = f'(x) .$$

As well, however, from the first moment that the differential functions as the starting point of the calculus, the inversion of the algebraic method of differentiation is complete, and the differential calculus itself therefore appears, a unique, specific method of calculating with variable quantities.

In order to make this more graphic I will combine at once all the algebraic methods which I have used, while setting simply $f(x)$ in place of a fixed algebraic expression in x, and the 'preliminary derivative' (see the first manuscript*) will be designated as $f^1(x)$ to distinguish it from the definitive 'derivative', $f'(x)$. Then, if

$$f(x) = y, \qquad f(x_1) = y_1,$$

[then]

$$f(x_1) - f(x) = y_1 - y \text{ or } \Delta y,$$

$$f^1(x)\,(x_1 - x) = y_1 - y \text{ or } \Delta y.$$

The preliminary derivative *must*† contain expressions in x_1

---

* See 'On the Concept of the Derived Function', p.3 above — *Ed.*

† The draft has: '*must* as a *rule*' — *Ed.*

and x exactly like the factor $(x_1 - x)$ with the *single exception* when $f(x)$ is an original function *to the first power*:

$$f^1(x) = \frac{y_1 - y}{x_1 - x} \text{ or } \frac{\Delta y}{\Delta x} .$$

We now substitute into $f^1(x)$

$$x_1 = x \text{ so that } x_1 - x = 0 ,$$

and thus obtain:

$$f'(x) = \frac{0}{0} \text{ or } \frac{dy}{dx}$$

and finally

$$f'(x)dx = dy \quad \text{or} \quad dy = f'(x)dx .$$

The differential of y is therefore the conclusion of an algebraic development; it becomes the starting point for differential calculus operating on its own ground. dy, the differential[26] of y — considered in isolation, that is, without its [real] equivalent — here immediately plays the same role as Δy in the algebraic method; and the differential of x, dx, the same role as Δx does there.

If we had, in

$$\frac{\Delta y}{\Delta x} = f^1(x)$$

cleared the denominator, then

I) $\Delta y = f^1(x)\,\Delta x$.

On the other hand, beginning with the differential calculus as a separate, complete type of calculating — and this point of departure has been itself derived algebraically — we start immediately with the differential expression of I), namely:

II) $dy = f'(x)dx$.

3) Since the symbolic differential equation (*Gleichung des Differentials*) arises simply by the algebraic handling of the most elementary functions with only one independent variable,

it appears that the inversion of the method (*Umschlag in der Methode*) could have been developed in a much more simple manner than happened with the example

$$y = uz\ .$$

The most elementary functions are those of the first degree; they are:

a) $y = x$, which leads to the differential coefficient $\frac{dy}{dx} = 1$, so that the differential is $dy = dx$.

b) $y = x \pm ab$; it leads to the differential coefficient $\frac{dy}{dx} = 1$, so that again the differential is $dy = dx$.

c) $y = ax$; it leads to the differential coefficient $\frac{dy}{dx} = a$, so that the differential is $dy = adx$.

Let us take the simplest case of all (under a)). Then:

$$y = x\ ,$$

$$y_1 = x_1\ ;$$

$$y_1 - y \text{ or } \Delta y = x_1 - x \text{ or } \Delta x\ .$$

I) $\frac{y_1 - y}{x_1 - x}$ or $\frac{\Delta y}{\Delta x} = 1$. Thus also $\Delta y = \Delta x$. In $\frac{\Delta y}{\Delta x}$ x_1 is now set $= x$, or $x_1 - x = 0$, and thus:

II) $\frac{0}{0}$ or $\frac{dy}{dx} = 1$; so that $dy = dx$.

Right at the start, as soon as we obtain I) $\frac{\Delta y}{\Delta x} = 1$, we are forced to operate further on the left-hand side, since on the right-hand side is the constant, 1. And therein the *inversion of the method*, which throws the initiative from the right-hand side to the left-hand side, once and for all from the ground up proves to be in fact the first word of the algebraic method itself.

Let's look at the matter more closely.

The real result was:

I) $\frac{\Delta y}{\Delta x} = 1$.

II) $\frac{0}{0}$ or $\frac{dy}{dx} = 1$.

Since both I) and II) lead to the same result we may choose between them. The setting of $x_1 - x = 0$ appears in any case to be a superfluous and therefore an arbitrary operation. Further: we operate from here on in II) on the left-hand side, since on the right-hand side 'ain't no way', so that we obtain:

$$\frac{0}{0} \text{ or } \frac{d^2y}{dx^2} = 0 .$$

The final conclusion would be that $\frac{0}{0} = 0$, so that the method is erroneous with which $\frac{0}{0}$ was obtained. At the first use* it leads to nothing new, and at the second to exactly nothing.[27]

Finally: we know from algebra that if the second sides of two equations are identical, so also must the first sides be. It therefore follows that:

$$\frac{dy}{dx} = \frac{\Delta y}{\Delta x} .$$

Since, however, both x and y, the variable dependent on x, are variable quantities, Δx while remaining a finite difference may be infinitely shortened; in other words it can *approach* 0 as closely as one wants, so that it becomes *infinitely small*; therefore the Δy dependent on it does so as well. Further, since $\frac{dy}{dx} = \frac{\Delta y}{\Delta x}$ it follows therefrom that $\frac{dy}{dx}$ really signifies, not the extravagant $\frac{0}{0}$, but rather the Sunday dress (*Sonntagsuniform*) of $\frac{\Delta y}{\Delta x}$, as soon as the latter functions as a ratio of infinitely small differences, hence differently from the usual difference calculation.

For its part the differential $dy = dx$ has no meaning, or more

* Original: '*coup*', French for 'strike', 'blow' — *Trans.*

correctly only as much meaning as we have discovered for both differentials in the analysis of $\frac{dy}{dx}$. Were we to accept the interpretation just given,[28] we could then perform miraculous operations with the differential, such as for example showing the role of adx in the determination of the subtangent of the parabola, which by no means requires that the nature of dx and dy really be understood.

4) Before I proceed to section III, which sketches the historical path of development of the differential calculus on an extremely condensed scale, here is one more example of the algebraic method applied previously. In order graphically to distinguish it I will place the given function on the left-hand side, which will always be the side of the initiative, since we always write from left to right, so that the general equation is:

$$x^m + Px^{m-1} + \text{ etc. } + Tx + U = 0 ,$$

and not

$$0 = x^m + Px^{m-1} + \text{ etc. } + Tx + U .$$

If the function y and the independent variable x are divided into two equations, of which the first expresses y as a function of the variable u, while on the other hand the second expresses u as a function of x, then *both symbolic differential coefficients in combination are to be found.*[29] Assuming:

1) $3u^2 = y ,\qquad 3u_1^2 = y_1 ,$

then

2) $x^3 + ax^2 = u ;\qquad x_1^3 + ax_1^2 = u_1 .$

We deal with equation 1) for the present:

$$3u_1^2 - 3u^2 = y_1 - y ,$$

$$3(u_1^2 - u^2) = y_1 - y ,$$

$$3(u_1 - u)(u_1 + u) = y_1 - y ,$$

$$3(u_1 + u) = \frac{y_1 - y}{u_1 - u} \quad \text{or} \quad \frac{\Delta y}{\Delta u} .$$

On the left-hand side u_1 is now set $= u$, so that $u_1 - u = 0$, then:

$$3(u + u) = \frac{dy}{du},$$

$$3(2u) = \frac{dy}{du},$$

$$6u = \frac{dy}{du}.$$

We now substitute for u its value $x^3 + ax^2$, so that:

3) $6(x^3 + ax^2) = \dfrac{dy}{du}$.

Now applying ourselves to equation 2):

$$x_1^3 + ax_1^2 - x^3 - ax^2 = u_1 - u,$$

$$(x_1^3 - x^3) + a(x_1^2 - x^2) = u_1 - u,$$

$$(x_1 - x)(x_1^2 + x_1x + x^2) + a(x_1 - x)(x_1 + x) = u_1 - u,$$

$$(x_1^2 + x_1x + x^2) + a(x_1 + x) = \frac{u_1 - u}{x_1 - x} \text{ or } \frac{\Delta u}{\Delta x}.$$

We set $x_1 = x$ on the left-hand side, so that $x_1 - x = 0$. Therefore

$$(x^2 + xx + x^2) + a(x + x) = \frac{du}{dx}.$$

4) $3x^2 + 2ax = \dfrac{du}{dx}$.

We now multiply equations 3) and 4) together, so that:

5) $6(x^3 + ax^2)(3x^2 + 2ax) = \dfrac{dy}{du} \cdot \dfrac{du}{dx} = \dfrac{dy}{dx}$.[30]

Thus, by algebraic means the operational formula

$$\frac{dy}{dx} = \frac{dy}{du} \cdot \frac{du}{dx},$$

has been found, which is also occasionally applicable to equations with two independent variables.

The above example shows that it is not witchcraft to transform a development demonstrated from given functions into a completely general form. Assume:

1) $y=f(u)$, $\quad y_1=f(u_1)$, $\quad y_1-y=f(u_1)-f(u)$,

so that $\quad$ therefore

2) $u=\varphi(x)$, $\quad u_1=\varphi(x_1)$, $\quad u_1-u=\varphi(x_1)-\varphi(x)$.

From the difference under 1) comes:

$$\frac{y_1-y}{u_1-u}=\frac{f(u_1)-f(u)}{u_1-u}\,; \qquad \frac{dy}{du}=\frac{df(u)}{du}\,,$$

however, since $df(u) = f'(u)du$,

$$\frac{dy}{du}=\frac{f'(u)du}{du}\,;$$

consequently

3) $\dfrac{dy}{du}=f'(u)$.

From the difference under 2) follows:

$$\frac{u_1-u}{x_1-x}=\frac{\varphi(x_1)-\varphi(x)}{x_1-x}\,, \qquad \frac{du}{dx}=\frac{d\varphi(x)}{dx}\,,$$

and since $d\varphi(x) = \varphi'(x)dx$,

$$\frac{du}{dx}=\frac{\varphi'(x)dx}{dx}\,,$$

so that:

4) $\dfrac{du}{dx}=\varphi'(x)$.

We multiply equation 3) by 4), so that:

5) $\frac{dy}{du} \cdot \frac{du}{dx}$ or $\frac{dy}{dx}$[31] $= f'(u) \,.\, \varphi'(x)$ Q.E.D.

N. III. The conclusion of this second instalment will follow, as soon as I consult John Landen at the Museum.[32]

Drafts and supplements on the work 'On the Differential'[33]

FIRST DRAFT[34]

As soon as we reach the differentiation of $f(u,z)$ $[=uz]$, where the variables u and z are both functions of x, we obtain — in contrast to the earlier cases which had only one *dependent variable*, namely, y — differential expressions on both sides, as follows:

in the *first instance*

$$\frac{dy}{dx} = z\frac{du}{dx} + u\frac{dz}{dx};$$

in the *second*, *reduced* form

$$dy = zdu + udz,$$

which last also has a form different from that in one dependent variable, as for example, $dy = max^{m-1}dx$, since here that immediately gives us the $\frac{dy}{dx}$ relieved of differential symbols $f'(x) = max^{m-1}$, which is by no means the case in $dy = zdu + udz$. The equations with one dependent variable showed us once and for all how the derived functions of [functions in] x, in this case of x^m, were obtained through actual differentiation [taking of differences] and their later cancellation, and at the same time how there arose the symbolic equivalent $\frac{0}{0} = \frac{dy}{dx}$ for the derived function. The substitution $\frac{0}{0} = \frac{dy}{dx}$ here appears not only permissible but even necessary, since $\frac{0}{0}$ in its primitive (*waldursprünglichen*) form = any quantity, because $\frac{0}{0} = X$ always leads to $0 = 0$. $\frac{0}{0}$ appears here, however, equal to an exactly defined (*ganz bestimmten*) specific value, $= mx^{m-1}$, and is itself the symbolic result of the operations whereby this value is derived from x^m; it is expressed as such a result in $\frac{dy}{dx}$. Thus $\frac{dy}{dx} \left(= \frac{0}{0}\right)$ is established from its origin as the symbolic value or differential expression of the already derived $f'(x)$, not, conversely, $f'(x)$ obtained by means of the symbol $\frac{dy}{dx}$.

At the same time, however, as soon as we have achieved this result and we therefore already operate on the ground (*Boden*) of differential calculus, we can reverse [the process]; if, for example, we have

$$x^m = f(x) = y$$

to differentiate, we know immediately (*von vornherein*)

$$dy = mx^{m-1}\,dx$$

or

$$\frac{dy}{dx} = mx^{m-1}.$$

Thus here we begin with the symbol; it no longer figures as the result of a derivation from the function [of] x; rather instead as already a *symbolic expression*[35] which indicates which operations to perform upon $f(x)$ in order to obtain the real value of $\frac{dy}{dx}$, i.e. $f'(x)$. In the first case $\frac{0}{0}$ or $\frac{dy}{dx}$ is obtained as the symbolic equivalent of $f'(x)$; and this is necessarily first, in order to reveal the origin of $\frac{dy}{dx}$; in the second case $f'(x)$ is obtained as the real value of the symbol $\frac{dy}{dx}$. But then, where the symbols $\frac{dy}{dx}$, $\frac{d^2y}{dx^2}$ become the operational formulae (*Operationsformeln*) of differential calculus,[36] they may as such formulae also appear on *the right-hand side of the equation*, as was already the case in the simplest example $dy = f'(x)dx$. If such an equation in its final form does not immediately give us, as in this case, $\frac{dy}{dx} = f'(x)$, etc., then this is proof that it is an equation which simply expresses symbolically which operations are to be performed in application to *defined* (*bestimmten*) functions.

And this is the case — and the simplest possible case — immediately in $d(uz)$, where u and z are both variables while both are also functions of the same third variable, i.e. of x.[37]

Given to be differentiated $f(x)$ or $y = uz$, where u and z are *both variables* dependent on x. Then

$$y_1 = u_1 z_1$$

and

$$y_1 - y = u_1 z_1 - uz\ .$$

Thus:

$$\frac{y_1 - y}{x_1 - x} = \frac{u_1 z_1}{x_1 - x} - \frac{uz}{x_1 - x}$$

or

$$\frac{\Delta y}{\Delta x} = \frac{u_1 z_1 - uz}{x_1 - x}\ .$$

But

$$u_1 z_1 - uz = z_1(u_1 - u) + u(z_1 - z)\ ,$$

since this is equivalent to

$$z_1 u_1 - z_1 u + uz_1 - uz = z_1 u_1 - uz\ .$$

Therefore:

$$\frac{u_1 z_1 - uz}{x_1 - x} = z_1 \frac{u_1 - u}{x_1 - x} + u \frac{z_1 - z}{x_1 - x}\ .$$

If now on both sides $x_1 - x$ becomes $= 0$, or $x_1 = x$, then we would have $u_1 - u = 0$, so that $u_1 = u$, and $z_1 - z = 0$, so that $z_1 = z$; we therefore obtain

$$\frac{dy}{dx} = z\frac{du}{dx} + u\frac{dz}{dx}$$

and therefore

$$d(uz) \text{ or } dy = zdu + udz\ .$$

At this point one may note in this differentiation of uz — in distinction to our earlier cases, where we had *only one dependent variable* that here we immediately find differential symbols on both sides of the equation, namely:

in the *first instance*

$$\frac{dy}{dx} = z\frac{du}{dx} + u\frac{dz}{dx}\ ;$$

in the *second*

$$d(uz) \text{ or } dy = zdu + udz$$

which also has a different form from that with one independent variable, such as for example, $dy = f'(x)dx$; for here division by dx immediately gives us $\frac{dy}{dx} = f'(x)dx$ which contains the specific value (*Spezialwert*) free of symbolic coefficients, derived from any function of x, $f'(x)$: which is in no sense the case in $dy = zdu + udz$.

It has been shown how, in functions with *only one independent variable*, from one function of x, for example $f(x) = x^m$, a second function of $x, f'(x)$, or, in the given case mx^{m-1} may be derived *by means of actual differentiation and subsequent cancellation alone*, and at the same time how from this process the symbolic equivalent $\frac{0}{0} = \frac{dy}{dx}$ for the derived function originates on the left-hand side of the equation.

Further: the substitution $\frac{0}{0} = \frac{dy}{dx}$ here was not only permissible but mathematically necessary. Since $\frac{0}{0}$ in its own primitive form may have any magnitude at all, for $\frac{0}{0} = X$ always gives $0 = 0$. Here, however, $\frac{0}{0}$ appears as the symbolic equivalent of a completely defined real value, as above, for example, mx^{m-1}, and is itself only the result of the operations whereby this value was derived from x^m; as such a result it is firmly fixed (*festgehalten*) in the form $\frac{dy}{dx}$.

Here, therefore, where $\frac{dy}{dx}\left(=\frac{0}{0}\right)$ is established in its origin, $f'(x)$ is by no means found by using the symbol $\frac{dy}{dx}$; rather instead the differential expression $\frac{dy}{dx}$ [appears] as the symbolic equivalent of the already derived function of x.

Once we have obtained this result, however, we can proceed in reverse. Given an $f(x)$, e.g. x^m, to differentiate, we then first look for the value of dy and find $dy = mx^{m-1}dx$, so that $\frac{dy}{dx} = mx^{m-1}$. Here the symbolic expression appears (*figuriert*) as the point of departure. [We] are thus (*so*) already operating on the ground of differential calculus; that is, $\frac{dy}{dx}$ etc. already perform as *formulae* which indicate which known differential operations to apply to the function of x. In the first case $\frac{dy}{dx}\left(=\frac{0}{0}\right)$ was obtained as the symbolic equivalent of $f'(x)$, in the second $f'(x)$ was sought and obtained as the real value of the symbols $\frac{dy}{dx}$, $\frac{d^2y}{dx^2}$, etc.

These symbols having already served as operational formulae (*Operationsformeln*) of differential calculus, they may then also appear on the right-hand side of the equation, as already happened in the simplest case, $dy = f'(x)dx$. If such an equation in its final form is not immediately reducible, as in the case mentioned, to $\frac{dy}{dx} = f'(x)$, that is to a real value, then that is proof that it is an equation which merely expresses symbolically which operations to use as soon as *defined functions* are treated in place of their undefined [symbols].

The simplest case where this comes in is $d(uz)$, where u and z are both variables, but both at the same time are functions of the same 3rd variable, e.g. of x.

If we have here obtained by means of the process of differentiation (*Differenzierungsprozess*) (see the beginning of this in Book I, repeated on p.10 of this book*)

$$\frac{dy}{dx} = x\frac{du}{dx} + u\frac{dz}{dx},$$

then we should not forget that u and z are here both *variables*, *dependent on* x, so y is only dependent on x, because on u and z. Where with *one* dependent variable we had it on the symbolic side, we now have the two variables u and z on the right-hand side, both independent with respect to y but both *dependent on* x, and their character [as] variables dependent on x appears in their respective symbolic coefficients $\frac{du}{dx}$ and $\frac{dz}{dx}$. If we deal with dependent variables on the right-hand side, then we must necessarily also deal with the differential coefficients on that side.

From the equation

$$\frac{dy}{dx} = z\frac{du}{dx} + u\frac{dz}{dx}$$

it follows:

$$d(uz) \text{ or } dy = zdu + udz.$$

This equation only indicates, however, the operations to perform when (*sobald*) u and z are given as defined functions.

The simplest possible case would be, for example,

$$u = ax\,, \qquad z = bx\,.$$

* See p.39 of this volume.

Then

$$d(uz) \quad \text{or} \quad dy = bx\,.\,adx + ax\,.\,bdx\ .$$

We divide both sides by dx, so that:

$$\frac{dy}{dx} = abx + bax = 2abx$$

and

$$\frac{d^2y}{dx^2} = ab + ba = 2ab\ .$$

If we take, however, the product from the very beginning,

$$y \quad \text{or} \quad uz = ax\,.\,bx = abx^2\,,$$

then

$$uz \quad \text{or} \quad y = abx^2\,, \qquad \frac{dy}{dx} = 2abx\ , \qquad \frac{d^2y}{dx^2} = 2ab\ .$$

As soon as we obtain a formula such as, for example, $[w =]\ z\,\frac{du}{dx}$, it is clear that the equation, 'what we might call'* a general operational equation, [is] a symbolic expression of the differential operation to be performed. If for example we take [the] expression $y\,\frac{dx}{dy}$, where y is the ordinate and x the abscissa, then this is the general symbolic expression for the subtangent of an arbitrary curve (exactly as $d(uz) = zdu + udz$ is the same for the differentiation of the product of two variables which themselves depend on a third). So long, however, as we leave the expression as it is it leads to nothing further, although we have the meaningful representation for dx, that it is the differential of the abscissa, and for dy, that it is the differential of the ordinate.

In order to obtain any positive result we must first take the equation of a definite curve, which gives us a definite value for y in x and therefore for dx as well, such as, for example,

* In English in original text — *Trans.*

$y^2 = ax$, the equation of the usual parabola; and then by means of differentiation we obtain $2ydy = adx$; hence $dx = \frac{2ydy}{a}$. If we substitute this definite value for dx into the general formula for the subtangent, $y\frac{dx}{dy}$, we then obtain

$$\frac{y\frac{2ydy}{a}}{dy} = \frac{y\,.\,2ydy}{ady} = \frac{2y^2}{a},$$

and since $y^2 = ax$, [this]

$$= \frac{2ax}{a} = 2x\,,$$

which is the value of the subtangent of the usual parabola; that is, it *is* $= 2 \times$ *the abscissa*. If, however, we call the subtangent τ, so that the general equation runs $y\frac{dx}{dy} = \tau$, and $ydx = \tau\,dy$. From the standpoint of the differential calculus, therefore, the question is usually (with the exception of Lagrange) posed thus: to find the real value for $\frac{dy}{dx}$.

The difficulty becomes evident if we then substitute the original form $\frac{0}{0}$ for $\frac{dy}{dx}$ etc.

$$\frac{dy}{dx} = z\frac{du}{dx} + u\frac{dz}{dx}$$

appears as

$$\frac{0}{0} = z\,.\,\frac{0}{0} + u\,.\,\frac{0}{0},$$

an equation which is correct but leads nowhere (*zu nichts*), all the less so, since the three $\frac{0}{0}$'s come from different differential coefficients whose different derivations are no longer visible. But consider:

1) Even in the first exposition with one independent variable, we first obtain

$$\frac{0}{0} \text{ or } \frac{dy}{dx} = f'(x)\ ; \qquad \text{so that } dy = f'(x)dx\ .$$

But since

$$\frac{dy}{dx} = \frac{0}{0}, \qquad dy = 0 \text{ and } dx = 0, \text{ so that } 0 = 0\ .$$

Although we again substitute for $\frac{dy}{dx}$ its indefinite expression $\frac{0}{0}$ we nonetheless commit here a positive mistake, for $\frac{0}{0}$ is only found here as the symbolic equivalent of the real value $f'(x)$, and as such is fixed in the expression $\frac{dy}{dx}$, and thus in $dy = f'(x)dx$ as well.

2) $\frac{u_1 - u}{x_1 - x}$ becomes $\frac{du}{dx}$ or $\frac{0}{0}$, because the variable x becomes $= x_1$, or $x_1 - x = 0$; we thus obtain right away not 0 but rather $\frac{0}{0}$ for $\frac{u_1 - u}{x_1 - x}$; we know however in general that $\frac{0}{0}$ can have any value, and that in a specific case it has the specific value (*Spezialwert*) which appears as soon as a defined function of x enters for u; we are thus not only correct in substituting $\frac{du}{dx}$ for $\frac{0}{0}$, but rather we must do it, since $\frac{du}{dx}$ as well as $\frac{dz}{dx}$ appear here only as symbols for the differential operations to be performed. So long as we stop with the result

$$\frac{dy}{dx} = z\frac{du}{dx} + u\frac{dz}{dx},$$

so that

$$dy = zdu + udz\ ,$$

then $\frac{du}{dx}$, $\frac{dz}{dx}$, *du* and *dz* also remain indefinite values, just like $\frac{0}{0}$ capable of any value.

3) In the usual algebra $\frac{0}{0}$ can appear as the form for expressions which have a real value, even though $\frac{0}{0}$ can be a symbol for any quantity. For example, given $\frac{x^2 - a^2}{x - a}$, we set $x = a$ so that $x - a = 0$ and $x^2 = a^2$, and therefore $x^2 - a^2 = 0$. We thus obtain

$$\frac{x^2 - a^2}{x - a} = \frac{0}{0};$$

the result so far is correct; but since $\frac{0}{0}$ may have any value it in no way proves that $\frac{x^2 - a^2}{x - a}$ has no real value.

If we resolve $x^2 - a^2$ into its factors, then it $= (x + a)(x - a)$; so that

$$\frac{x^2 - a^2}{x - a} = (x + a) . \frac{x - a}{x - a} = x + a ;$$

so if $x - a = 0$, then $x = a$, so therefore $x + a = a + a = 2a$.[38]

If we had the term $P(x - a)$ in an ordinary algebraic equation, then if $x = a$, so that $x - a = 0$, then necessarily $P(x - a) = P . 0 = 0$; just as under the same assumptions $P(x^2 - a^2) = 0$. The decomposition of $x^2 - a^2$ into its factors $(x + a)(x - a)$ would change none of this, for

$$P(x + a)(x - a) = P(x + a) . 0 = 0 .$$

By no means, however, does it therefore follow that if the term $P . \left(\frac{0}{0}\right)$ had been developed by setting $x = a$, its value must necessarily be $= 0$.

$\frac{0}{0}$ may have any value because $\frac{0}{0} = X$ always leads to: $0 = X . 0 = 0$; but just because $\frac{0}{0}$ may have any value it need not necessarily have the value 0, and if we are acquainted with its origin we are also able to discover a real value hidden behind it.

So for example $P.\frac{x^2-a^2}{x-a}$, if $x = a$, $x-a = 0$ and so as well $x^2 = a^2$, $x^2-a^2 = 0$; thus

$$P.\frac{x^2-a^2}{x-a} = P.\frac{0}{0}.$$

Although we have obtained this result in a mathematically completely correct manner, it would nonetheless be mathematically false, however, to conclude without further ado that $P.\frac{0}{0} = 0$, because such an assumption would imply that $\frac{0}{0}$ may necessarily have no value other than 0, so that

$$P.\frac{0}{0} = P.0.$$

It would be more relevant to investigate whether any other result arises from resolving x^2-a^2 into its factors, $(x+a)(x-a)$; in fact, this transforms the expression to

$$P.(x+a).\frac{x-a}{x-a} = P.(x+a).1,$$

and [when] $x = a$ to $P.2a$ or $2Pa$. Therefore, as soon as we operate (*rechnen*) with variables,[39] it is all the more not only legitimate but indeed advisable to fix firmly (*festzuhalten*) the origin of $\frac{0}{0}$ by the use of the differential symbols $\frac{dy}{dx}$, $\frac{dz}{dx}$, etc., after we have previously (*ursprünglich*) proved that they originate as the symbolic equivalent of derived functions of the variables which have run through a definite process of differentiation. If they are thus originally (*ursprünglich*) the result of such a process of differentiation, then they may for that reason well become *inversely* (*umgekehrt*) symbols of a process yet to be performed on the variables, thus *operational symbols*

(*Operationssymbolen*) which appear as points of departure rather than results, and this is their essential use (*Dienst*) in differential calculus. As such operational symbols they may even convey the contents of the equations among the different variables (in implicit functions 0 stands from the very beginning on the right-hand side [of the equation] and the dependent as well as independent variables, together with their coefficients, on the left).

Thus it is in the equation which we obtain:

$$\frac{d(uz)}{dx} \quad \text{or} \quad \frac{dy}{dx} = \frac{zdu}{dx} + \frac{udz}{dx}.$$

From what has been said earlier it may be observed that the dependent functions of x, z and u, here appear unchanged as z and u again; but each of them is equipped (*ausgestattet*) with the factor of the symbolic differential coefficient of the other.

The equation therefore only has the value of a general equation which indicates by means of symbols which operations to perform as soon as u and z are given respectively, as dependent variables, two defined functions of x.

Only when [we] have defined functions of [x] for u and z may $\frac{du}{dx}\left(=\frac{0}{0}\right)$ and $\frac{dz}{dx}\left(=\frac{0}{0}\right)$ and therefore $\frac{dy}{dx}\left(=\frac{0}{0}\right)$ as well become 0, so that the value $\frac{0}{0} = 0$ cannot be presumed but on the contrary must have arisen from the defined functional equation itself.

Let, for example, $u = x^3 + ax^2$; then

$$\left(\frac{0}{0}\right) = \frac{du}{dx} = 3x^2 + 2ax\,,$$

$$\left(\frac{0}{0}\right)_1 = \frac{d^2u}{dx^2} = 6x + 2a\,,$$

$$\left(\frac{0}{0}\right)_2 = \frac{d^3u}{dx^3} = 6\,,$$

$$\left(\frac{0}{0}\right)_3 = \frac{d^4u}{dx^4} = 0\,,$$

so that in this case $\frac{0}{0} = 0$.

The long and the short of the story is that here by means of differentiation itself we obtain the *differential coefficients in their symbolic form as a result*, as the value of $\left[\frac{dy}{dx}\text{ in}\right]$ the differential equation, namely in the equation

$$\frac{d(uz)}{dx} \quad \text{or} \quad \frac{dy}{dx} = z\frac{du}{dx} + u\frac{dz}{dx}.$$

We now know, however, that $u =$ a defined function of x, say $f(x)$. Therefore $\frac{u_1 - u}{x_1 - x}$, in its differential symbol $\frac{du}{dx}$, is equal to $f'(x)$, the first derived function of $f(x)$. Just so $z = \varphi(x)$, say, and so similarly $\frac{dz}{dx} = \varphi'(x)$, ditto — of $\varphi(x)$. The original function itself, however, provides us neither with u nor with z in any defined function of x, such as, for example

$$u = x^{\mathrm{m}}, \quad z = \sqrt{x}$$

It provides us u and z only as general expressions for any 2 arbitrary functions of x whose product is to be differentiated.

The equation states that, if a product, represented by uz, of any two functions of x is to be differentiated, one is first to find the real value corresponding to the symbolic differential coefficient $\frac{du}{dx}$, that is the first derived function say of $f(x)$, and to multiply this value by $\varphi(x) = z$; then similarly to find the real value of $\frac{dz}{dx}$ and multiply [it] by $f(x) = u$; and finally to add the two products thus obtained. The operations of differential calculus are here already assumed to be well-known.

The equation is thus only a symbolic indication of the operations to be performed, and at the same time the symbolic differential coefficients $\frac{du}{dx}$, $\frac{dz}{dx}$ here stand for symbols of differential operations still to be completed in any concrete case, while they themselves were originally derived as symbolic formulae for already completed differential operations.

As soon as they have taken on (*angenommen*) this character, they may themselves become the contents of differential equation, as, for example, in *Taylor's Theorem*:

$$y_1 = y + \frac{dy}{dx} h + \text{etc.}$$

But then these are also only general, symbolic operational equations. In this case of the differentiation of uz, the interest lies in the fact that it is the simplest case in which — in distinction to the development of those cases where the independent variable x has only one dependent variable y — differential symbols due to the application of the original method itself are placed as well on the right-hand side of the equation (its developed expression), so that at the same time they enter as operational symbols and as such became the contents of the equation itself.

This role, in which they indicate operations to be performed and therefore serve as the point of departure, is their characteristic role in a differential calculus already operating (*sich bewegenden*) on its own ground, but it is certain (*sicher*) that no mathematician has taken account of this inversion, this reversal of roles, still less has it been necessary to demonstrate it using a totally elementary differential equation. It has only been mentioned as a matter of fact that, while the discoverers of the differential calculus and the major part of their followers make the differential symbol the point of departure for calculus, Lagrange in reverse makes the algebraic derivation[40] of actual (*wirklichen*) functions of the independent variable the point of departure, and the differential symbols into merely symbolic expressions of already derived functions.

If we once more return to $d(uz)$, we obtained next as the result (*Produkt*) of setting $x_1 - x = 0$, as the result of the differential operation itself:

$$\frac{dy}{dx} = z\,\frac{du}{dx} + u\,\frac{dz}{dx}\,.$$

Since there is a common denominator here, we thus obtain as a reduced expression

$$dy = z du + u dz\,.$$

This compares to (*entspricht*) the fact that in the case of only one dependent variable we obtain as the symbolic expression of the derived function of x, of $f'(x)$ (for instance, of max^{m-1}, which is $f'(x)$ if $ax^m = f(x)$), $\frac{dy}{dx}$ on the left-hand side as its symbolic expression

$$\frac{dy}{dx} = f'(x)$$

and of which the first result is

$$dy = f'(x)\,dx$$

(for example, $\frac{dy}{dx} = max^{m-1}$; $dy = max^{m-1}dx$, which is the differential of the function y) (which last we may equally re-transform to $\frac{dy}{dx} = max^{m-1}$). But the case

$$dy = z du + u dz$$

is distinguished once again by reason of the fact that the differentials du, dz here lie on the right-hand side, as operational symbols, and that dy is only defined after the completion of the operations which they indicate. If

$$u = f(x)\ ,\quad z = \varphi(x)$$

then we know that we obtain for du

$$du = f'(x)\,dx$$

and for [dz]

$$dz = \varphi'(x)dx$$

Therefore:

$$dy = \varphi(x)f'(x)\,dx + f(x)\,\varphi'(x)\,dx$$

and

$$\frac{dy}{dx} = \varphi(x)f'(x) + f(x)\,\varphi'(x)\ .$$

In the first case therefore first the differential coefficient

$$\frac{dy}{dx} = f'(x)$$

is found and then the differential

$$dy = f'(x)dx\ .$$

In the second case first the differential dy and then the differential coefficient $\frac{dy}{dx}$. In the first case, where the differential symbols themselves first originate from the operations performed with $f(x)$, first the derived function, the true (*wirkliche*) differential coefficient, must be found, to which $\frac{dy}{dx}$ stands opposite (*gegenübertrete*) as its symbolic expression; and only after it has been found can the differential (*das Differential*) $dy = f'(x)dx$ be derived.

It is turned round (*umgekehrt*) in $dy = zdu + udz$.

Since du, dz appear here as operational symbols and clearly indicate operations which we already know, from differential calculus, how to carry out, therefore we must first, in order to find the real value of $\frac{dy}{dx}$, in every concrete case substitute for u its value in x, and for z ditto—its value in x—in order to find

$$dy = \varphi(x)f'(x)\,dx + f(x)\,\varphi'(x)\,dx;$$

and then for the first time division by dx provides the real value of

$$\frac{dy}{dx} = \varphi(x)f'(x) + f(x)\,\varphi'(x)\ .$$

What is true for $\frac{du}{dx}$, $\frac{dz}{dx}$, $\frac{dy}{dx}$, $\frac{d^2y}{dx^2}$ etc. is true for all complicated formulae where *differential symbols* themselves appear within general symbolic operational equations.

SECOND DRAFT[41]

[I]

. . . .

We start with the algebraic derivation of $f'(x)$, in order to establish in this way at the same time its symbolic differential expressions $\frac{0}{0}$ or $\frac{dy}{dx}$, and thus also discover its meaning. We must then turn it round, starting with the symbolic differential coefficients $\frac{du}{dx}$, $\frac{dz}{dx}$ as given forms in order to find their respective corresponding real equivalents $f'(x)$, $\varphi'(x)$. And indeed, these different ways of treating the differential calculus, setting out from opposite poles — and two different historical schools — here do not arise from changes in our subjective methods, but from the nature of the function *uz* to be dealt with. We deal with it, as with functions of *x* with a single dependent variable, by starting with the right-hand pole and operating algebraically with it. I do not believe any mathematician has proved or rather even noticed this necessary reversal from the first method of algebraic derivation (historically the second) whether for so elementary a function as *uz* or any other. They were too absorbed with the material of the calculus for this.

Indeed, we find that in the equation

$$\frac{0}{0} \quad \text{or} \quad \frac{dy}{dx} = z\frac{du}{dx} + u\frac{dz}{dx}$$

$\frac{dy}{dx}$ again springs in just the same way from the derivative occurring on the right, with *uz* just as with functions of *x* with a single dependent variable; but on the other hand the differential symbols $\frac{du}{dx}$, $\frac{dz}{dx}$ are again incorporated in $f'(x)$ or the first derivative of uz, and therefore form elements of the equivalent of $\frac{dy}{dx}$.

The symbolic differential coefficients thus themselves become already the *object* or *content* of the differential operation, instead of as before featuring as its purely symbolic result (*als symbolisches Resultat derselben*).

With these two points, *first,* that the symbolic differential coefficients as well as the variables become substantial elements of the derivation, become *objects* of differential operations (*Differentialoperationen*), *second*, that the question has changed about, from finding the symbolic expression for the real differential coefficient $f'(x)$, to finding the real differential coefficient for its symbolic expression — with both these points the third is given, that instead of appearing as the symbolic result of the previous operation of differentiation on the real function of x, the symbolic differential expressions now conversely (*umgekehrt*) play the role of symbols which indicate operations of differentiation yet to be performed on the real function of x; that they thus become *operational symbols*.

In our case, where

$$\frac{dy}{dx} = z\frac{du}{dx} + u\frac{dz}{dx},$$

we would no longer be able to operate unless we knew not only that z and u are both functions of x but also that, just as with

$$y = x^{m},$$

real values in x are given for u and z, such as, for example,

$$u = \sqrt{x}, \qquad z = x^3 + 2ax^2.$$

In that manner, then, $\frac{du}{dx}$, $\frac{dz}{dx}$ in fact stand as indicators of operations whose performance (*Ausführungsweise*) is assumed to be well-known for any arbitrary function of x substituted in place of u and z.

c) The equation found is not only a symbolic operational equation (*Operationsgleichung*), but also simply a preparatory symbolic operational equation.

Since in

$$\text{[I)]} \qquad \frac{dy}{dx} = z\frac{du}{dx} + u\frac{dz}{dx},$$

the denominator dx is found in all terms on both sides, its reduced expression is thus:

II) dy or $d(uz) = zdu + udz$.

Straight away this equation says that when a product of two arbitrary variables (and this is generalisable in further applications to the product of an arbitrary number of variables) is to be differentiated, each of the two factors is to be multiplied by the differential of the other factor and the two products so obtained are to be added.

The first operational equation

$$\frac{dy}{dx} = z\frac{du}{dx} + u\frac{dz}{dx}$$

thus becomes, if the product of two arbitrary variables is to be differentiated, a superfluous preparatory equation which, after it has served its purpose, namely that of a general symbolic operational formula, leads directly to the goal.

And here it may be remarked that the process of the original algebraic derivation is again turned into its opposite. We first obtained there

$$\Delta y = y_1 - y$$

as the corresponding symbol for $f(x_1) - f(x)$, both usual algebraic expressions (since $f(x)$ and $f(x_1)$ have been given as defined algebraic functions of x). Then $\frac{f(x_1) - f(x)}{x_1 - x}$ was replaced by $\frac{\Delta y}{\Delta x}$, whereupon $f'(x)$ — the first derived function of $f(x)$ — became $\frac{dy}{dx}$, and we at last obtained, from the final equation of the differential coefficient,

$$\frac{dy}{dx} = f'(x) \ ,$$

the differential

$$dy = f'(x)dx \ .$$

The above equation,* however, gives us the differentials dy, du, dz as points of departure (*Ausgangspunkte*). Thus, were in fact arbitrary defined functions of x to be substituted for u and z, designated only as

$$u = f(x) \quad \text{and} \quad z = \varphi(x) \ ,$$

then we would have

$$dy = \varphi(x)\, df(x) + f(x)\, d\varphi(x) \ ,$$

and this d sign merely indicates differentiation to be performed.

The result of this differentiation has the general form:

$$df(x) = f'(x)\, dx$$

and

$$d\varphi(x) = \varphi'(x)\, dx \ .$$

So that

$$dy = \varphi(x) f'(x)\, dx + f(x)\varphi'(x)\, dx.$$

Finally,

$$\frac{dy}{dx} = \varphi(x) f'(x) + f(x)\varphi'(x).$$

Here, where the differential already plays the role of a ready-made operational symbol, we therefore derive the differential coefficient from it; while on the contrary in the original

* Equation II) — *Trans.*

algebraic development the differential was derived from the equation for the differential coefficient.

Let us take the *differential* itself, as we have developed it in its simplest form, namely, from the function of the first degree:

$$y = ax \; , \qquad \frac{dy}{dx} = a \; ;$$

of which the differential is

$$dy = adx \; .$$

The equation of this differential appears to be much more meaningful than that of the differential coefficient,

$$\frac{0}{0} \text{ or } \frac{dy}{dx} = a \; ,$$

from which it is derived.

Since $dy = 0$ and $dx = 0$, $dy = adx$ is identical to $0 = 0$. Yet, we are completely correct to use dy and dx for the vanished — but fixed, by means of these symbols, in their disappearance — differences, $y_1 - y$ and $x_1 - x$.

As long as we stay with the expression

$$dy = adx$$

or, in general,

$$dy = f'(x)\, dx \; ,$$

it is nothing other than a restatement of the fact that

$$\frac{dy}{dx} = f'(x) \; ,$$

which in the above case, $= a$, from which we may continue to transform it further. But this ability to be transformed already makes it an operational symbol (*Operationssymbole*). At once, we see that if we have found $dy = f'(x)\, dx$ as a result of the process of differentiation, we have only to divide both sides by dx to find $\frac{dy}{dx} = f'(x)$, namely, the differential coefficient.

Thus for example in $y^2 = ax$

$$d(y^2) = d(ax) \ , \qquad 2ydy = adx \ .$$

The last equation of differentials provides us with two equations of differential coefficients, namely:

$$\frac{dy}{dx} = \frac{a}{2y} \quad \text{and} \quad \frac{dx}{dy} = \frac{2y}{a} \ .$$

But $2ydy = adx$ also provides us *immediately* with the value $\frac{2ydy}{a}$ for dx, which for instance substitutes into the general formula for the subtangent $y\frac{dx}{dy}$ and finally helps to establish $2x$, double the abscissa, as the value of the subtangent of the usual parabola.

II

We now want to take an example in which these symbolic expressions first serve the calculus as ready-made (*fertige*) operational formulae, so that the real value of the symbolic coefficient is also found and then the reversed elementary algebraic exposition may be followed.

1) The dependent function y and the independent variable x are not united in a single equation, but in such a manner that y appears in a first equation as a direct function of the variable u, and then u in a second equation as a direct function of the variable x. The task: *to find the real value of the symbolic differential coefficient,* $\frac{dy}{dx}$.

Let

$$\text{a) } y = f(u) \ , \qquad \text{b) } u = \varphi(x) \ .$$

Next, 1) $y = f(u)$ gives:

$$\frac{dy}{du} = \frac{df(u)}{du} = \frac{f'(u)du}{du} = f'(u) \ .$$

2) $\frac{du}{dx} = \frac{d\varphi(x)}{dx} = \frac{\varphi'(x)dx}{dx} = \varphi'(x)$.

So that

$$\frac{dy}{du} \cdot \frac{du}{dx} = f'(u) . \varphi'(x) .$$

But

$$\frac{dy}{du} \cdot \frac{du}{dx} = \frac{dy}{dx} .$$

So that

$$\frac{dy}{dx} = f'(u)\varphi'(x) .$$

Example. If a) $y = 3u^2$, b) $u = x^3 + ax^2$, then, by the formula

$$\frac{dy}{du} = \frac{d(3u^2)}{du} = 6u \ (= f'(u)) ;$$

but the equation b) says $u = x^3 + ax^2$. If we substitute this value for u in $6u$, then

$$\frac{dy}{du} = 6(x^3 + ax^2) \quad (=f'(u)) .$$

Furthermore:

$$\frac{du}{dx} = 3x^2 + 2ax \quad (=\varphi'(x)) .$$

So that

$$\frac{dy}{du} \cdot \frac{du}{dx} \text{ or } \frac{dy}{dx} = 6(x^3 + ax^2)(3x^2 + 2ax) \ (=f'(u) . \varphi'(x)) .$$

2) We now take the equations contained in the last example as the starting equations (*Ausgangsgleichungen*), in order to develop them this time in the first, algebraic, method.

a) $y = 3u^2$,   b) $u = x^3 + ax^2$.
Since $y = 3u^2$, [then] $y_1 = 3u_1^2$, and

$$y_1 - y = 3(u_1^2 - u^2) = 3(u_1 - u)\,(u_1 + u) .$$

Therefore

$$\frac{y_1 - y}{u_1 - u} = 3(u_1 + u) .$$

If now $u_1 - u$ becomes $= 0$, then $u_1 = u$, and $3(u_1 + u)$ is thus transformed to $3(u + u) = 6u$.

We substitute for u its value in equation b), so that

$$\frac{dy}{du} = 6(x^3 + ax^2) .$$

Further; since

$$u = x^3 + ax^2 , \quad \text{[then]} \quad u_1 = x_1^3 + ax_1^2 ;$$

so that

$$u_1 - u = (x_1^3 + ax_1^2) - (x^3 + ax^2) = (x_1^3 - x^3) + a(x_1^2 - x^2) ,$$

$$u_1 - u = (x_1 - x)\,(x_1^2 + x_1x + x^2) + a(x_1 - x)\,(x_1 + x) ;$$

thus

$$\frac{u_1 - u}{x_1 - x} = (x_1^2 + x_1x + x^2) + a(x_1 + x) .$$

If now $x_1 - x$ becomes $= 0$ then $x_1 = x$, so that

$$x_1^2 + x_1x + x^2 = 3x^2$$

and

$$a(x_1 + x) = 2ax .$$

Thus:

$$\frac{du}{dx} = 3x^2 + 2ax .$$

If we now multiply both equations together, we then obtain on the right-hand side

$$6(x^3 + ax^2)\,(3x^2 + 2ax)\ ,$$

which corresponds to the left-hand side

$$\frac{dy}{du} \cdot \frac{du}{dx} = \frac{dy}{dx}\ ,$$

just as previously.

In order to bring out the difference in the derivations more clearly, we shall place the defined functions of the variables on the left-hand side and the functions dependent on them on the right-hand side, since one is accustomed, following the general equations in which only 0 stands on the right hand, to thinking that the initiative is on the left-hand side. Thus:

a) $3u^2 = y$; b) $x^3 + ax^2 = u$.

Since

$$3u^2 = y\ , \qquad 3u_1^2 = y_1\ ,$$

so that

$$3(u_1^2 - u^2) = y_1 - y$$

or

$$3(u_1 - u)\,(u_1 + u) = y_1 - y\ ,$$

so that

$$3(u_1 + u) = \frac{y_1 - y}{u_1 - u}\ .$$

If now u_1 becomes $= u$, so that $u_1 - u = 0$, [we] then obtain

$$3(u + u) \text{ or } 6u = \frac{dy}{du}\ .$$

If we substitute in $6u$ its value from equation b), then

$$6(x^3 + ax^2) = \frac{dy}{du}\ .$$

Furthermore, if

$$x^3 + ax^2 = u ,$$

then

$$x_1^3 + ax_1^2 = u_1$$

and

$$x_1^3 + ax_1^2 - x^3 - ax^2 = u_1 - u ;$$

so that

$$(x_1^3 - x^3) + a(x_1^2 - x^2) = u_1 - u .$$

We further separate into factors:

$$(x_1 - x)\,(x_1^2 + x_1x + x^2) + a(x_1 - x)\,(x_1 + x) = u_1 - u .$$

Therefore

$$(x_1^2 + x_1x + x^2) + a(x_1 + x) = \frac{u_1 - u}{x_1 - x} ;$$

now if $x_1 = x$, so that $x_1 - x = 0$, then

$$3x^2 + 2ax = \frac{du}{dx} .$$

If we multiply the 2 derived functions together, then

$$6(x^3 + ax^2)\,(3x^2 + 2ax) = \frac{dy}{dx} ,$$

and if we put it in the usual order,

$$\frac{dy}{du} \cdot \frac{du}{dx} = \frac{dy}{dx} = 6(x^3 + ax^2)\,(3x^2 + 2ax) .$$

It is self-evident that due to its details and the frequently difficult division of the first difference, $f(x_1) - f(x)$, into terms which each contain the factor $x_1 - x$, the latter method is not comparable to the historically older one as a means of calculation.

On the other hand one begins this last method with dy, dx and $\frac{dy}{dx}$ as given operational formulae, while one sees them arise in the first one, and in a purely algebraic manner as well. And I maintain nothing more. And there in the [historically] first method, how has the point of departure of the differential symbols as operational formulae been obtained? Either through covertly or through overtly metaphysical assumptions, which themselves lead once more to metaphysical, unmathematical consequences, and so it is at that point that the violent suppression is made certain, the derivation is made to start its way, and indeed quantities made to proceed from themselves.

And now, in order to give an historical example of beginning from the two opposing poles, I will compare the solution of the case of $d(uz)$ developed above by Newton and Leibnitz on the one hand, to that by Lagrange on the other hand.

1) *Newton*.

We are first told that when the variable quantities increase $\dot{x}$, $\dot{y}$ etc. designate the velocities of their fluxions, *alias* of the increase, respectively, of x, y etc. Since furthermore the numerical sizes of all possible quantities may be represented by means of straight lines, the *momenta* or *infinitely small quanta* which are produced are equal to the *product* of the velocities $\dot{x}$, $\dot{y}$ etc. with the infinitely small time intervals τ during which they occur, thus $= \dot{u}\tau$, $\dot{x}\tau$ and $\dot{y}\tau$.[42]

'THIRD DRAFT'

If we now consider the differential of y in its general form, $dy = f'(x)dx$, then we already have before us a purely symbolic operational equation, even in the case where $f'(x)$ from the very beginning is a constant, as in $dy = d(ax) = adx$. This child of $\frac{0}{0}$ or $\frac{dy}{dx} = f'(x)$ looks suspiciously like its mother. For in $\frac{dy}{dx} = \frac{0}{0}$ *numerator and denominator are inseparably bound together*; in $dy = f'(x)dx$ they are obviously separated, so that one is forced to the conclusion: $dy = f'(x)dx$ is only a masked expression for $0 = f'(x) . 0$, thus $0 = 0$ with which 'nothing's to be done' (*'nichts zu wolle'*). Looking more closely, analysts in our century, such as, for example, the Frenchman Boucharlat, smell a rat here too. He says:★

'In $\frac{dy}{dx} = 3x^2$, for example, $\frac{0}{0}$ alias $\frac{dy}{dx}$, or even more its value $3x^2$, is the differential coefficient of the function y. Since $\frac{dy}{dx}$ is thus the symbol which represents the limit $3x^2$, dx must always stand under dy but, in order *to facilitate algebraic operation* we treat $\frac{dy}{dx}$ as an ordinary fraction and $\frac{dy}{dx} = 3x^2$ as an ordinary equation, and thus by removing the denominator dx from the equation obtain the result $dy = 3x^2dx$, which expression is called *the differential of y*.'[43]

In order to 'facilitate algebraic operation', we thus introduce a false formula.

★ This is a translation of Marx's German translation of a passage originally in French — *Trans.*

In fact the thing (*Sache*) doesn't behave that way. In $\frac{0}{0}$ (usually written $\left(\frac{0}{0}\right)$), the ratio of the minimal expression (*Minimalausdrucks*) of $y_1 - y$, or of $f(x_1) - f(x)$, or of the increment of $f(x)$, to the minimal expression of $x_1 - x$, or to the increment of the independent variable quantity x, possesses a form in which the numerator is inseparable from the denominator. But why? In order to retain $\frac{0}{0}$ as the *ratio* of vanished differences. As soon, however, as $x_1 - x = 0$ obtains in dx a form which manifests it as the vanished difference of x, and thus $y_1 - y = 0$ appears as dy as well, the separation of numerator and denominator becomes a completely permissible operation. Where dx now stands its relationship with dy remains undisturbed by this change of position. $dy = df(x)$, and thus $= f'(x)\,dx$, is only another expression for $\frac{dy}{dx}\left[= f'(x)\right]$, which must lead to the conclusion that $f'(x)$ is obtained independently. How useful this formula $dy = df(x)$ immediately becomes as an *operational formula* (*Operationsformel*), however, is shown, for example, by:

$$y^2 = ax\ ,$$

$$d(y^2) = d(ax)\ ,$$

$$2ydy = adx\ ;$$

so that

$$dx = \frac{2ydy}{a}\ .$$

This value of dx, substituted into the general formula for the subtangent, $y\frac{dx}{dy}$, then gives

$$\frac{y\frac{2ydy}{a}}{dy} = \frac{2y^2dy}{ady} = \frac{2y^2}{a},$$

and since

$$y^2 = ax\,,\ [\text{thus}]\ = \frac{2ax}{a} = 2x\,;$$

so that $2x$, double the abscissa, is the value of the subtangent of the usual parabola.

However, if $dy = df(x)$ serves as the first point of departure (*Ausgangspunkt*), which only later is developed into $\frac{dy}{dx}$ itself, then, for this differential of y to have any meaning at all, these differentials dy, dx must be *assumed* to be symbols with a defined meaning. Had such assumptions not originated from mathematical metaphysics but instead been derived quite directly from a function of the first degree, such as $y = ax$, then, as seen earlier, this leads to $\frac{y_1 - y}{x_1 - x} = a$, which is transformed to $\frac{dy}{dx} = a$. From here as well, however, nothing certain is to be got *a priori*. For since $\frac{\Delta y}{\Delta x}$ is just as much $= a$ as $\frac{dy}{dx} = a$, and the Δx, Δy, although finite differences or increments, are yet finite differences or increments of unlimited capacity to contract (*Kontraktionsfähigkeit*), one then may just as well represent dx, dy as infinitely small quantities capable of arbitrarily approaching 0, as if they originate from actually setting the equality $x_1 - x = 0$, and thus as well $y_1 - y = 0$. The result remains identical on the right-hand side in both cases, because there in itself there is no x_1 at all to set $= x$, and thus as well no $x_1 - x = 0$. This substitution $= 0$ on the other side consequently appears just as arbitrary an hypothesis as the assumption that dx, dy are arbitrarily small quantities. Under (*sub*) IV) I will briefly indicate the historical development through the example of $d(uz)$, but yet prior to that will give an example under (*sub*) III)[44] which is treated the first time on the ground of symbolic calculus, with a ready-made operational formula (*fertigen Operationsformel*), and is demonstrated a second time algebraically. Enough (*soviel*) has been shown under (*sub*) II), so that the latter method alone, by means of its application to so elementary a function as the product of two variables, using its own results, necessarily leads to starting points (*die Ausgangspunkte*) which are the opposite pole as far as operating method goes.

To (*ad*) IV.

Finally (*following Lagrange*) it is to be noted that the *limit* or the *limit value*, which is already occasionally found in Newton for the differential coefficients and which he still derives from purely geometric considerations (*Vorstellungen*), still to this day always plays a predominant role, whether the symbolic expressions appear (*figurieren*) as the limit of $f'(x)$ or conversely $f'(x)$ appears as the limit of the symbol or the two appear together as limits. This category, which Lacroix in particular analytically broadened, only becomes important as a substitute for the category 'minimal expression', whether it is of the derivative as opposed to the 'preliminary derivative', or of the ratio $\frac{y_1 - y}{x_1 - x}$, when the application of calculus to curves is treated. It is *more representable* (*vorstellbarer*) geometrically and is already found therefore *among the old geometricians*. Some contemporaries (*Modernen*) still hide behind the statement that the differentials and differential coefficients merely express very approximate values.[45]

SOME SUPPLEMENTS[46]

A) Supplement on the differentiation of uz.[47]

1) For me the essential thing in the last manuscript on the development of $d(uz)$ was the proof, referring to the equation

$$\text{A)}\quad \frac{dy}{dx} = z\frac{du}{dx} + u\frac{dz}{dx},$$

that the algebraic method applied here reverses itself into the differential method, since it develops within the derivative, and thus on the right-hand side, *symbolic differential coefficients* without corresponding equivalents, real coefficients; hence these symbols as such become *independent starting points* and ready-made *operational formulae*.

The form of equation A) lends itself all the more readily to this purpose since it allows a comparison between the $\frac{du}{dx}, \frac{dz}{dx}$, produced within the derivative $f'(x)$, and the $\frac{dy}{dx}$, which is the symbolic differential coefficient of $f'(x)$ and therefore comprises its symbolic equivalent, standing opposite on the left-hand side.

Confronting the character of $\frac{du}{dx}, \frac{dz}{dx}$, as operational formulae, I have been content with the hint that for any symbolic differential coefficient an arbitrary 'derivative' may be found as its real value if one substitutes some $f(x)$, $3x^2$ for example, for u and some $\varphi(x)$, $x^3 + ax^2$ for example, for z.

I however could also have indicated the geometric applicability of each operational formula, since for example, *the general formula for the subtangent of a curve* $= y\frac{dx}{dy}$, which has a form generally identical to $z\frac{du}{dx}$, $u\frac{dz}{dx}$, for they are all products of a variable and a symbolic differential coefficient.

Finally, it could have been noted that $y = uz$ [is] the *simplest elementary function* (y here $= y^1$, and uz is the simplest form of the second power) with which our theme could have been developed.

A) Differentiation of $\frac{u}{z}$.[48]

3) Since $d\frac{u}{z}$ is the inverse of $d(uz)$, where one has multiplication, the other division, one may use the *algebraically* obtained operational formula

$$d(uz) = zdu + udz$$

in order to find $d\frac{u}{z}$ directly. I will now do this, in order that the difference between the method of derivation and the simple application of a differential result found previously which now in turn serves as an operational formula, may stand out clearly.

$$\text{a)}\ y = \frac{u}{z},$$

$$\text{b)}\ u = yz\ .$$

Since $y=\frac{u}{z}$, thus

$$yz = \frac{u}{z} . z = u\ .$$

We have thus simply formally concealed u in a product of two factors. Nonetheless, the task is thereby in fact already solved, since the problem has been transformed from the differentiation of a fraction to the differentiation of a product, for which we have the magic formula in our pocket. According to this formula:

$$\text{c)}\ du = zdy + ydz\ .$$

We see immediately that the first term of the second side, namely zdy, must remain *sitting in peace* at its post until the crack of doom (*genau vor Torschluss*), since the task consists precisely in finding the differential of $y\left(=\frac{u}{z}\right)$, and thus its expression in differentials of u and z. For this reason, on the other hand, ydz is to be *removed* to the left-hand side. Therefore:

$$\text{d)}\quad du - ydz = zdy\ .$$

We now substitute the value of y, namely $\frac{u}{z}$, into ydz, so that

$$du - \frac{u}{z}\,dz = zdy\ ;$$

therefore

$$\frac{zdu - udz}{z} = zdy\ .$$

The moment has now come to free dy of its 'sleeping partner'* z, and we obtain

$$\frac{zdu - udz}{z^2} = dy = d\frac{u}{z}\ .$$

---

* Original in English — *Trans.*

On the history of Differential Calculus[49]

A PAGE INCLUDED IN THE NOTEBOOK 'B (CONTINUATION OF A) II'[50]

1) *Newton*, born 1642, †1729. *'Philosophiae naturalis principia mathematica'*, pub. 1687.

L. I. *Lemma* XI, *Schol.* Lib. II.

L. II. *Lemma* II, *from Proposition* VII.[51]

'Analysis per quantitatum series, fluxiones etc.', composed 1665, publ. 1711.[52]

2) *Leibnitz.*

3) *Taylor* (J. Brook), born 1685, †1731, published 1715-17: *'Methodus incrementorum etc.'*

4) *MacLaurin* (*Colin*), born 1698, †1746.

5) *John Landen.*

6) *D'Alembert*, born 1717, †1783. *'Traité des fluides'*, 1744.[53]

7) *Euler* (Léonard), [born] 1707, †1783. *'Introduction à l'analyse de l'infini'*, Lausanne, 1748. *'Institutions du calcul différentiel'*, 1755 (p.I, c.III).[54]

8) *Lagrange*, born 1736. *'Théorie des fonctions analytiques'* (1797 and 1813) (*see Introduction*).

9) *Poisson* (Denis, Siméon), born 1781, †1840.

10) *Laplace* (P. Simon, marquis de), born 1749, †1827.

11) *Moigno's*, *'Leçons de Calcul Différentiel et de calcul intégral'*.[55]

I. FIRST DRAFTS

Newton: born 1642, †1727 (85 years old). *Philosophiae naturalis principia mathematica* (first published 1687; c.f. *Lemma* I and *Lemma* XI, *Schol.*)

Then in particular: *Analysis per quantitatum series fluxiones etc.*, first published 1711, but composed in 1665, while Leibnitz first made the same discovery in 1676.

Leibnitz: born 1646, †1716 (70 years old).

Lagrange: born 1736, †during the Empire (Napoleon I); he is the discoverer of the *method of variations*. *Théorie des fonctions analytiques* (1797 and 1813).

D'Alembert: born 1717, †1783 (66 years old). *Traité des fluides*, 1744.

1) *Newton*. The velocities or fluxions, of for example the *variables* x, y etc. *are denoted by* $\dot{x}, \dot{y}$ etc. For example if u and x are *connected quantities (fluents) generated by continuous movement*, then $\dot{u}$ and $\dot{x}$ denote their rates of increase, and therefore $\frac{\dot{u}}{\dot{x}}$ the *ratio* of the rates at which their increments are generated.

Since the numerical quantities of all possible magnitudes may be represented by straight lines, and the *moments* or infinitely small portions of the quantities generated = the products of their velocities and the infinitely small time intervals during which these velocities exist,[56] so then [we have] τ denoting these infinitely small time intervals, and the moments of x and y represented by $\tau\dot{x}$ and $\tau\dot{y}$, respectively.

For example: $y = uz$; [with] $\dot{y}, \dot{z}, \dot{u}$ denoting the velocities at which y, z, u respectively [are] increasing, then the *moments* of $\dot{y}, \dot{z}, \dot{u}$ are $\tau\dot{y}, \tau\dot{z}, \tau\dot{u}$, and we obtain

$$y = uz \,,$$

$$y + \tau\dot{y} = (u + \tau\dot{u})\,(z + \tau\dot{z}) = uz + u\tau\dot{z} + z\tau\dot{u} + \tau^2\dot{u}\dot{z} \,;$$

hence

$$\tau\dot{y} = u\tau\dot{z} + z\tau\dot{u} + \tau^2\dot{u}\dot{z} \,.$$

Since τ is infinitesimal, it disappears by itself and even more as the product $\tau^2\dot{u}\dot{z}$ altogether, since it is not that of the infinitely small period of time τ, but rather its 2nd power. $\left(\text{If } \tau = \frac{1}{\text{million}}, \text{ then } \tau^2 = \frac{1}{1 \text{ million} \times 1 \text{ million}}\right)$.

We thus obtain

$$\dot{y} = \dot{u}z + \dot{z}u \,,$$

or the fluxion of $y = uz$ is $\dot{u}z + \dot{z}u$.[57]

2) *Leibnitz*. The differential of uz is to be found.
u becomes $u + du$, z becomes $z + dz$; so that

$$uz + d(uz) = (u + du)\,(z + dz) = uz + udz + zdu + dudz \,.$$

If from this the given quantity uz is subtracted, then there remains $udz + zdu + dudz$ as the increment; $dudz$, the product *d'un infiniment petit du par un autre infiniment petit dz*, (of an infinitely small du times another infinitely small dz)* is an infinitesimal of the second order and disappears before the infinitesimals udz and zdu of the first order; therefore

$$d(uz) = udz + zdu \,.$$[58]

[3)] *D'Alembert*. Puts the problem in general terms thus: If [we have]

$$y = f(x) \,,$$

$$y_1 = f(x + h) \,;$$

* In French in the original. — *Trans.*

[we are] to determine what the value of $\frac{y_1 - y}{h}$ becomes when the quantity h disappears, and thus what is the value of $\frac{0}{0}$.[59]

Newton and Leibnitz, like the majority of the successors from the beginning performed operations on the ground of the differential calculus, and therefore valued differential expressions from the beginning as operational formulae whose real equivalent is to be found. All of their intelligence was concentrated on that. If the independent variable x goes to x_1, then the dependent variable goes to y_1. $x_1 - x$, however, is necessarily equal to some difference, let us say, $= h$. This is contained in the very concept of variables. In no way, however, does it follow from this that this difference, which $= dx$, is a vanished [quantity], so that in fact it $= 0$. It may represent a finite difference as well. If, however, we suppose from the very beginning that x, when it increases, goes to $x + \dot{x}$ (the τ which Newton uses serves no purpose in his analysis of the fundamental functions and so may be suppressed[60]), or, with Leibnitz, goes to $x + dx$, then differential expressions immediately become operational symbols (*Operationssymbole*) without their algebraic origin being evident.

To 15* (*Newton*).

Let us take Newton's beginning equation for the product uz that is to be differentiated; then:

$$y = uz \ ,$$

$$y + \tau\dot{y} = (u + \dot{u}\tau)\,(z + \dot{z}\tau) \ .$$

---

* See pages 49-51 in this edition.

If we toss out the τ, as he does himself if you please, after he develops the first differential equation, we then obtain:

$$y + \dot{y} = (u + \dot{u})\,(z + \dot{z})\ ,$$

$$y + \dot{y} = uz + \dot{u}z + \dot{z}u + \dot{z}\dot{u}\ ,$$

$$y + \dot{y} - uz = \dot{u}z + \dot{z}u + \dot{u}\dot{z}\ .$$

So that, since $uz = y$,

$$\dot{y} = \dot{u}z + \dot{z}u + \dot{u}\dot{z}\ .$$

And in order to obtain the correct result $\dot{u}\dot{z}$ must be suppressed.

Now, whence arises the term to be forcibly suppressed, $\dot{u}\dot{z}$?

Quite simply from the fact that the differentials $\dot{y}$ of y, $\dot{u}$ of u, and $\dot{z}$ of z have from the very beginning been imparted by definition* a separate, independent existence from the variable quantities from which they arose, without having been derived in any mathematical way at all.

On the one hand one sees what usefulness this presumed existence of dy, dx or $\dot{y}$, $\dot{x}$ has, since from the very beginning, as soon as the variables increase I have only to substitute in the algebraic function the binomials $y + \dot{y}$, $x + \dot{x}$ etc. and then may just manipulate (*manövrieren*) these themselves as ordinary algebraic quantities.

I obtain, for example, if I have $y = ax$:

$$y + \dot{y} = ax + a\dot{x}\ ;$$

so that

$$y - ax + \dot{y} = a\dot{x}\ ;$$

hence

$$\dot{y} = a\dot{x}\ .$$

I have therewith immediately obtained the result: the differential of the dependent variable is equal to the increment of

* Original: 'Difinition', presumably 'Definition' — *Trans.*

ax, namely $a\dot{x}$; it is equal to *the real value a derived* from ax* (that this is a constant quantity here is accidental and does nothing to alter the generality of the result, since it is due to the circumstance that the variable x appears here to the first power). If I generalise this result,[61] then I know $y = f(x)$, for this means that y is the variable dependent on x. If I call the quantity derived from $f(x)$, i.e. the real element of the increment, $f'(x)$, then the general result is:

$$\dot{y} = f'(x)\dot{x} .$$

I thus know from the very beginning that the equivalent of the differential of the dependent variable y is equal to the first derived function of the independent variable, multiplied by its differential, that is dx or $\dot{x}$.

So then, generally expressed, if

$$y = f(x)$$

then

$$dy = f'(x)dx$$

or $\dot{y}$ = the real coefficient in x (except where a constant appears because x is to the first power) times $\dot{x}$.

But $\dot{y} = a\dot{x}$ gives me immediately $\frac{\dot{y}}{\dot{x}} = a$, and in general:

$$\frac{\dot{y}}{\dot{x}} = f'(x) .$$

I have thus found for the differential and the differential coefficients two fully-developed operational formulae which form the basis of all of differential calculus.

And furthermore, put in general terms, I have obtained, by means of assuming dx, dy etc. or $\dot{x}$, $\dot{y}$ etc. to be independent, insulated increments of x and y, the enormous advantage, distinctive to the differential calculus, that all functions of the variables are expressed from the very beginning in differential forms.

---

* That is, $\frac{\dot{y}}{\dot{x}} = a$ – *Trans.*

Were I thus to develop the essential functions of the variables in this manner, such as ax, $ax \pm b$, xy, $\frac{x}{y}$, x^{n}, a^{x}, $\log x$, as well as the elementary trigonometric functions then the determination of dy, $\frac{dy}{dx}$ would thus become completely tamed, like the multiplication table in arithmetic.

If we now look, however, on the reverse side we find immediately that the entire original operation is mathematically false.

Let us take a perfectly simple example: $y = x^2$. If x increases then it contains an indeterminate increment h, and the variable y dependent on it has an indeterminate increment k, and we obtain

$$y + k = (x + h)^2 = x^2 + 2hx + h^2 \ ,$$

a formula which is given to us by the binom[ial theorem]. Therefore

$$y + k - x^2 \quad \text{or} \quad y + k - y = 2hx + h^2\,;$$

hence

$$(y + k) - y \quad \text{or} \quad k = 2hx + h^2\,;$$

if we divide both sides by h then:

$$\frac{k}{h} = 2x + h \ .$$

We now set $h = 0$, and this becomes

$$2x + h = 2x + 0 = 2x \ .$$

On the other side, however, $\frac{k}{h}$ goes to $\frac{k}{0}$. Since, however, y only went to $y + k$ because x went to $x + h$, and then $y + k$ goes back to y when h goes to 0, therefore when $x + h$ goes back to $x + 0$, to x. So then k also goes to 0 and $\frac{k}{0} = \frac{0}{0}$, which may be expressed as $\frac{dy}{dx}$ or $\frac{\dot{y}}{\dot{x}}$. We thus obtain:

$$\frac{0}{0} \text{ or } \frac{\dot{y}}{\dot{x}} = 2x .$$

If on the other hand we [substitute $h = 0$] in

$$y + k - x^2 = 2hx + h^2 \quad \text{or} \quad (y + k) - y = 2xh + h^2$$

(h is only replaced by the symbol dx after it has previously been set equal to 0 in its original form), we then obtain $k = 0 + 0 = 0$, and the sole result that we have reached is the insight into our assumption, merely that y goes to $y + k$, if x goes to $x + h$ . . . so that if $x + h = x + 0 = x$, then $y + k = y$, or $k = 0$.

In no way do we obtain what Newton makes of it:

$$k = 2x dx + dx dx$$

or, in Newton's way of writing:

$$\dot{y} = 2x\dot{x} + \dot{x}\dot{x} ;$$

h only becomes $\dot{x}$, and therefore k becomes $\dot{y}$, as soon as h has passed the hellish ride through 0, that is, subsequent to the difference $x_1 - x$ (or $(x + h) - x$) and therefore that of $y_1 - y$ as well ($= (y + k) - y$) having been reduced to their absolutely minimum expressions (*Mimimalausdruck*), $x - x = 0$ and $y - y = 0$.

Since Newton, however, does not immediately determine the increments of the variables x, y, etc by means of mathematical derivation, but instead immediately stamps $\dot{x}$, $\dot{y}$, etc on to the differentials, they cannot be set $= 0$; for otherwise, were the result 0, which is algebraically expressed as setting this increment from the very beginning $= 0$, it would follow from that, just as above in the equation

$$(y + k) - y = 2xh + h^2 ,$$

h would immediately be set equal to 0, therefore $k = 0$, and consequently in the final analysis we would obtain $0 = 0$. The nullification of h may not take place prior to the first derived function of x, here $2x$, having been freed of the factor h through division, thus:

$$\frac{y_1 - y}{h} = 2x + h\ .$$

Only then may the finite differences be annulled. The differential coefficient

$$\frac{dy}{dx} = 2x$$

therefore also must have previously been developed,[62] before we may obtain the differential

$$dy = 2x dx$$

Therefore nothing more remains than to imagine the increments h of the variable to be infinitely small increments and to give them as such *independent existence*, in the symbols $\dot{x}$, $\dot{y}$ etc. or dx, dy [etc.] for example. But infinitely small quantities are quantities just like those which are infinitely large (the word infinitely (*unendlich*) [small] only means in fact indefinitely (*unbestimmt*) small); the dy, dx etc. or $\dot{y}$, $\dot{x}$ [etc.] therefore also take part in the calculation just like ordinary algebraic quantities, and in the equation above

$$(y + k) - y \quad \text{or} \quad k = 2x dx + dx dx$$

the $dx dx$ has the same right to existence as $2x dx$ does; the reasoning (*Raisonnement*) is therefore most peculiar by which it is forcibly suppressed, namely, by direct use of the relativity of the concept of infinitesimal (*unendlich klein*). $dx dx$ is suppressed because it is infinitely small compared to dx, and thus as well to $2x dx$, or to $2x\dot{x}$. . .

But (*Oder*), if in

$$\dot{y} = \dot{u}z + \dot{z}u + \dot{u}\dot{z}$$

the $\dot{u}\dot{z}$ is suppressed because it is infinitely small compared to $\dot{u}z$ or $\dot{z}u$, then one would thereby be forced to admit mathematically that $\dot{u}z + \dot{z}u$ is only an approximate value (*Annäherungswert*), in imagination as close as you like. This type of manoeuvre occurs also in ordinary algebra.

But then in walks the still greater miracle that by this method you don't obtain an approximate value at all, but rather the unique exact value (even when as above it is only symbolically correct) of the derived function, such as in the example

$$\dot{y} = 2x\dot{x} + \dot{x}\dot{x} \ .$$

If you suppress here $\dot{x}\dot{x}$, you then obtain:

$$\dot{y} = 2x\dot{x}$$

and

$$\frac{\dot{y}}{\dot{x}} = 2x \ ,$$

which is the correct first derived function of x^2, as the binom[ial theorem] has already proved.

But the miracle is no miracle. It would only be a miracle if *no exact result* emerged through the forcible suppression of $\dot{x}\dot{x}$. That is to say, one *suppresses merely a computational mistake* which nevertheless is an unavoidable consequence of a method which brings in the undefined increment of the variable, i.e. h, immediately as the differential dx or $\dot{x}$, a completed operational symbol, and thereby also produces from the very beginning in the differential calculus a characteristic manner of calculation different from the usual algebra.

The general direction of the algebraic method which we have applied may be expressed as follows:

Given $f(x)$, first develop the 'preliminary derivative', which we would like to call $f^1(x)$:

1) $$f^1(x) = \frac{\Delta y}{\Delta x} \quad \text{or} \quad \frac{\Delta y}{\Delta x} = f^1(x) \ .$$

From this equation it follows

$$\Delta y = f^1(x)\Delta x \ .$$

So that as well

$$\Delta f(x) = f^1(x)\Delta x$$

(since $y = f(x)$, [thus] $\Delta y = \Delta f(x)$) .

By means of setting $x_1 - x = 0$, so that $y_1 - y = 0$ as well, we obtain

[2)] $$\frac{dy}{dx} = f'(x) .$$

Then

$$dy = f'(x)dx ;$$

so that also

$$df(x) = f'(x)dx$$

(since $y = f(x)$, $dy = df(x)$) .

When we have once developed

1) $\Delta f(x) = f^1(x)\Delta x$,

then

2) $df(x) = f'(x)dx$

is only the differential expression of 1).

1) If we have x going to x_1, then

A) $x_1 - x = \Delta x$;

whence the following conclusions may be drawn

Aa) $\Delta x = x_1 - x$; a) $x_1 - \Delta x = x$;

Δx, the *difference* between x_1 and x, is therefore positively expressed as the *increment* of x; for when it is subtracted again from x_1 the latter returns once more to its original state, to x.

The difference may therefore be expressed in two ways: *directly as the difference* between the increased variable and its state before the increase, and this is its *negative expression*; positively as the increment,* *as a result*: *as the increment* of x to the state in which it has not yet grown, and this is the positive expression.

We shall see how this double formulation plays a role in the history of differential calculus.

b) $x_1 = x + \Delta x$.

x_1 is the increased x itself; its growth is not separated from it; x_1 is the completely indeterminate form of its growth. This formula distinguishes the increased x, namely x_1, from its original form prior to the increase, from x, but it does not distinguish x from its own increment. The relationship between x_1 and x may therefore only be expressed negatively, as a *difference*, as $x_1 - x$. In contrast, in

$$x_1 = x + \Delta x$$

1) The difference is expressed *positively* as an increment of x.

2) Its increase is therefore not expressed as a *difference*, but instead as the *sum* of itself in its original state + its increment.

3) Technically x is expelled from its monomial into a binomial, and wherever x appears to any power in the original function a binomial composed of *itself and its increment* enters for the increased x; the binomial $(x + h)^m$ in general for x^m. The development of the increase of x is therefore in fact a simple application of the *binomial theorem*. Since x enters as the first and Δx as the second term of this binomial — which is given by their very relationship, since x must be [there] before the formation of its increment Δx — by means of the binomial,

* Marx added here in pencil 'or decrement' — *Ed.*

in the event only the functions of x will be derived, while Δx figures next to it as a factor raised to increasing powers; indeed, Δx to the first power must ⌈appear⌉, so that Δx^1 is a factor of the second term of the resulting series, of the first function, that is, of x_1 derived, using the binomial theorem. This shows up perfectly when x is given to the second power. x^2 goes to $(x+h)^2$, which is nothing more than the *multiplication* of $x+\Delta x$ by itself, ⌈and which⌉ leads to $x^2 + 2x\Delta x + \Delta x^2$: that is, the first term must be the original function of x and the first derived function of x^2, namely ⌈2⌉x here, comprises the second term together with the factor Δx^1, which entered into the first term only as the factor $\Delta x^0 = 1$. So then, the derivative is not found by means of differentiation but rather by means of the application of the binomial theorem, therefore *multiplication*; and this because the increased variable x_1 takes part from the very beginning as a binomial, $x+\Delta x$.

4) Although Δx in $x+\Delta x$ is just as indefinite, so far as its magnitude goes, as the indefinite variable x itself, Δx is defined as a distinct quantity, separate from x, like a fruit beside the mother who had previously borne her (*als Frucht neben ihrer Mutter, bevor diese geschwangert war*).

$x+\Delta x$ not only expresses in an indefinite way the fact that x has increased as a variable; rather, it ⌈also⌉ expresses *by how much* it has grown, namely, by Δx.

5) x never appears as x_1; the whole development centres around the increment Δx as soon as the derivative has been found by means of the binomal theorem, by means, that is, of substituting $x+\Delta x$ for x in a definite way (*in bestimmten Grad*). On the left-hand side, however, if in $\frac{y_1 - y}{\Delta x}$, the Δx becomes $= 0$, it finally appears as $x_1 - x$ again, so that:

$$\frac{y_1 - y}{\Delta x} = \frac{y_1 - y}{x_1 - x}\,^{\star}.$$

★ Marx added here in pencil: $\frac{\Delta y}{\Delta x}$. — *Ed.*

The positive side, where $x_1 - x = 0$ takes place, namely x_1 becoming $= x$, can therefore never enter into the development, since x_1 as such never enters into the side of the resultant series (*Entwicklungsreihe*); the real mystery of the differential calculus makes itself evident as never before.

6) If $y = f(x)$ and $y_1 = f(x+\Delta x)$, then we can say that in using this method the *development of* y_1 *solves the problem of finding the derivative*.

c) $x+\Delta x = x_1$ (so that $y+\Delta y = y_1$ as well). Δx here may only appear in the form $\Delta x = x_1 - x$, therefore in the *negative* form of the *difference* between x_1 and x, and not in the *positive* form of the increment of x, as in $x_1 = x+\Delta x$.

1) Here the increased x is distinguished as x_1 *from itself*, before it grows, namely from x; but x_1 does not appear as an x increased by Δx, so x_1 therefore remains just exactly as indefinite as x is.

2) Furthermore: however x enters into any original function, so x_1 does as the increased variable in the original function now altered by the increase. For example, if x takes part in the function x^3, so does x_1 in the function x_1^3.

Whereas previously, by means of substituting $(x+\Delta x)$ wherever x appeared in the original function, the derivative had been provided ready-made by the use of the binomial, leaving it burdened with the factor Δx and the first of other terms in x burdened with Δx^2 etc., so now there is just as little which can be derived directly from the immediate form of the monomial — x_1^3 — as could be got from x^3. It does provide, however, the *difference* $x_1^3 - x^3$. We know from algebra that all differences of the form $x^3 - a^3$ are divisible by $x - a$; the given case, therefore, is divisible by $x_1 - x$. In therefore dividing $x_1^3 - x^3$ by $x_1 - x$ (instead of, [as] previously, multiplying the term $(x+\Delta x)$ by itself to the degree specified by the function), we obtain an expression of the form $(x_1 - x)P$, wherein nothing is affected whether the original function of x contains many terms (and so contains x to various powers) or as in our example is of a single term. This $x_1 - x$ passes by division to the denominator of $y_1 - y$ on the left-hand side and thus produces $\frac{y_1 - y}{x_1 - x}$ there, the ratio of the difference of the function to the difference of the independent variable x in its abstract difference-formula (*Differenzform*). The development of the difference between the function expressed in x_1 and that expressed in x into terms, all of which have $x_1 - x$ as a factor, may well require algebraic manipulation (*Manöver*) to a greater or lesser degree, and thus may not always shed as much light as the form $x_1^3 - x^3$. This has no effect on the method.

When by its nature the original function does not allow the direct development into $(x_1 - x)P$, as was the case with $f(x) = uz$ (two variables both dependent on x), $(x_1 - x)$ appears [in] the factor $\frac{1}{x_1 - x}$. Furthermore, after the removal of $x_1 - x$ to the left-hand side by means of dividing both sides by it, $x_1 - x$ still continues to exist in P itself (as, for example, in the derivation from $y = a^x$, where we find

$$\frac{y_1 - y}{x_1 - x} = a^x \left\{(a - 1) + \frac{(x_1 - x) - 1}{1.2}(a - 1)^2 + \text{ etc.}\right\},$$

where setting $x_1 - x = 0$ produces

$$= a^x \left\{(a - 1) - \frac{1}{2}(a - 1)^2 + \frac{1}{3}(a - 1)^3 - \text{ etc.}\right\});$$

this is only possible when, as in the example just given, it so happens that setting $x_1 - x = 0$ [allows] it to disappear, and then always leaves positive results behind in its place. In other words the $(x_1 - x)$s left behind in P may not be combined with the rest of the elements of P as factors (as multiplicands). P would otherwise be factorable into $P = p(x_1 - x)$, and then, since $x_1 - x$ has already been set $= 0$, into $p.0$; hence $P = 0$ . . .[63]

The first finite difference, $x_1^3 - x^3$, where $y = x^3$ and $y_1 = x_1^3$, has therefore been evolved to

$$y_1 - y = (x_1 - x)P\ ,$$

hence

$$\frac{y_1 - y}{x_1 - x} = P\ .$$

P, an expression combining x_1 and x, is $= f^1$, the derivative of the first finite difference, whence $x_1 - x$ has been quite eliminated, as well as those of higher degree, $(x_1 - x)^2$ etc. x_1 and x may therefore only be combined in positive expressions, such as $x_1 + x$, $x_1 x$, $\frac{x_1}{x}$, $\sqrt{x_1 x}$ etc. Were therefore x_1 to be now set $= x$, these expressions would then become $2x$, x^2, $\frac{x}{x}$ or 1, $\sqrt{xx}$ or x etc., respectively, and only on the left-hand side, where $x_1 - x$ comprises the denominator, is 0 produced and therefore the symbolic differential coefficient etc.

II. THE HISTORICAL PATH OF DEVELOPMENT

1) Mystical Differential Calculus. $x_1 = x + \Delta x$ from the beginning changes into $x_1 = x + dx$ or $x + \dot{x}$, where dx is assumed by metaphysical *explanation*. First it exists, and then it is explained.

Then, however, $y_1 = y + dy$ or $y_1 = y + \dot{y}$. From the arbitrary assumption the consequence follows that in the expansion of the binomial $x + \Delta x$ or $x + \dot{x}$, the terms in x and Δx which are obtained in addition to the first derivative, for instance, must be *juggled away* in order to obtain the correct result etc. etc. Since the real foundation of the differential calculus proceeds from this last result, namely from the *differentials* which anticipate and are not derived but instead are *assumed* by explanation, then $\frac{dy}{dx}$ or $\frac{\dot{y}}{\dot{x}}$ as well, the symbolic differential coefficient, is *anticipated* by this explanation.

If the increment of $x = \Delta x$ and the increment of the variable dependent on it $= \Delta y$, then it is self-evident (*versteht sich von selbst*) that $\frac{\Delta y}{\Delta x}$ represents the ratio of the increments of x and y. This implies, however, that Δx figures in the denominator, that is the increase of the independent variable is in the denominator instead of the numerator, not the reverse; while the final result of the development of the differential form, namely *the differential*, is also given in the very beginning by the assumed differentials.*

---

* Marx distinguishes the differentials (*die Differentiellen*) dx and dy, the infinitesimals of the differences Δx and Δy, from the differential (*das Differential*): $dy = f'(x)dx$. — *Trans.*

If I assume the simplest possible (*allereinfachste*) ratio of the dependent variable y to the independent variable x, then $y = x$. Then I know that $dy = dx$ or $\dot{y} = \dot{x}$. Since, however, I seek the derivative of the independent [variable] x, which here $= \dot{x}$, I therefore have to divide[64] both sides by $\dot{x}$ or dx, so that:

$$\frac{dy}{dx} \quad \text{or} \quad \frac{\dot{y}}{\dot{x}} = 1 .$$

I therefore know once and for all that in the symbolic differential coefficient the increment [of the independent variable] must be placed in the denominator and not in the numerator.

Beginning, however, with functions of x in the second degree, the *derivative* is found immediately by means of the binomial theorem [which provides an expansion] where it appears ready made (*fix und fertig*) in the second term combined with dx or $\dot{x}$; that is with the increment of the first degree + the terms to be juggled away. The *sleight of hand* (*Eskamotage*), however, is unwittingly mathematically correct, because it only juggles away errors of calculation arising from the original sleight-of-hand in the very beginning.

$x_1 = x + \Delta x$ is to be changed to

$$x_1 = x + dx \quad \text{or} \quad x + \dot{x} ,$$

whence this differential binomial may then be treated as are the usual binomials, which from the technical standpoint would be very convenient.

The only question which still could be raised: why the mysterious suppression of the terms standing in the way? That specifically assumes that one knows they stand in the way and do not truly belong to the derivative.

The *answer* is very simple: this is found purely by experiment. Not only have the true derivatives been known for a long time, both of many more complicated functions of x as well as of their analytic forms as equations of curves, etc., but they have also been discovered by means of the most decisive experiment possible, namely by the treatment of the simplest algebraic function of second degree for example:

$$y = x^2$$
$$y + dy = (x + dx)^2 = x^2 + 2x\,dx + dx^2 ,$$
$$y + \dot{y} = (x + \dot{x})^2 = x^2 + 2x\dot{x} + \dot{x}^2 .$$

If we subtract the original function, $x^2 (y = x^2)$, from both sides, then:

$$dy = 2x\;dx + dx^2$$
$$\dot{y} = 2x\dot{x} + \dot{x}\dot{x} ;$$

I suppress the last terms on both [right] sides, then:

$$dy = 2x\;dx , \qquad \dot{y} = 2x\dot{x} ,$$

and further

$$\frac{dy}{dx} = 2x ,$$

or

$$\frac{\dot{y}}{\dot{x}} = 2x .$$

We know, however, that the first term out of $(x + a)^2$ is x^2; the second $2xa$; if I divide this expression by a, as above $2x\,dx$ by dx or $2x\dot{x}$ by $\dot{x}$, we then obtain $2x$ as the first derivative of x^2, namely the increase in x,[65] which the binomial has added to x^2. Therefore the dx^2 or $\dot{x}\dot{x}$ had to be suppressed in order to find the derivative; completely neglecting the fact that nothing could begin with dx^2 or $\dot{x}\dot{x}$ * in themselves.

In the experimental method, therefore, one comes — right at the second step — necessarily to the insight that dx^2 or $\dot{x}\dot{x}$ has to be juggled away, not only to obtain the true result but any result at all.

Secondly, however, we had in

$$2x\;dx + dx^2 \quad \text{or} \quad 2x\dot{x} + \dot{x}\dot{x}$$

* Printed edition has misprint $x\dot{x}$ here. — *Trans.*

the true mathematical expression (second and third terms) of the binomial $(x + dx)^2$ or $(x + \dot{x})^2$. That *this mathematically correct result* rests on *the mathematically basically false assumption* that $x_1 - x = \Delta x$ is from the beginning $x_1 - x = dx$ or $\dot{x}$, was not known.[66]

In other words, instead of using sleight of hand, one obtained the same result by means of an algebraic operation of the simplest kind and presented it to the mathematical world.

Therefore: mathematicians (*man . . . selbst*) really believed in the mysterious character of the newly-discovered means of calculation which led to the correct (and, particularly in the geometric application, surprising) result by means of a positively false mathematical procedure. In this manner they became themselves mystified, rated the new discovery all the more highly, enraged all the more greatly the crowd of old orthodox mathematicians, and elicited the shrieks of hostility which echoed even in the world of non-specialists and which were necessary for the blazing of this new path.

2) *Rational Differential Calculus*. D'Alembert starts directly from the *point de départ* (*sic*) of Newton and Leibnitz: $x_1 = x + dx$. But he immediately makes the fundamental correction: $x_1 = x + \Delta x$, that is, x and an *undefined* but prima facie *finite increment* which he calls h. The transformation of this h or Δx into dx (he uses the Leibnitz notation, like all Frenchmen) is first found as the final result of the development or at least just before the gate swings shut (*vor Toresschluss*), while in the mystics and the initiators of the calculus it appears as the starting point (d'Alembert himself begins with the symbolic side,* but first transforms it symbolically). By this means he immediately succeeds in two ways.[67]

a) The ratio of differences

$$\frac{f(x+h) - f(x)}{h} = \frac{f(x+h) - f(x)}{x_1 - x}$$

is the starting point of his construction (*Bildung*).

* Traditionally the left-hand side — *Trans.*

1) [the difference] $f(x+h)-f(x)$, corresponding to the given algebraic function in x, stands out as soon as you replace x itself with its increment $x+h$ in the original function in x, for example, in x^3. This form ($=y_1-y$, if $y=f(x)$) is that of the *difference of the function*, whose transformation into a ratio of the increment of the function to the increment of the independent variable now requires a development, so that it plays a real role instead of a merely nominal one, as it does with the mystics; for, if I have in these authors

$$f(x) = x^3 ,$$

$$f(x+h) = (x+h)^3 = x^3 + 3x^2h + 3xh^2 + h^3 ,$$

then I know from the very beginning, that in

$$f(x+h)-f(x) = x^3 + 3x^2h + 3xh^2 + h^3 - x^3 ,$$

the opposing sides are to be reduced to the increment. This needn't even be written down, since I see that on the second side the increment of x^3 = the three following terms as well as that in $f(x+h)-f(x)$, only the increment of $f(x)$ remains, or dy. The first difference equation (*Differenzgleichung*) therefore only plays a role which from the very beginning is to disappear again. The increments stand opposite one another on both sides, and if I have them then I have from the definition of dx, dy that $\frac{dy}{dx}$ or $\frac{\dot{y}}{\dot{x}}$ is the ratio etc.; I therefore do not need the first difference, formed by the subtraction of the original function in x from the altered (by the replacement of x by $x+h$) function (the increased function), in order to construct $\frac{dy}{dx}$ or $\frac{\dot{y}}{\dot{x}}$.

In d'Alembert it is necessary to hold fast to this difference because the steps of the development (*Entwicklungsbewegungen*) are to be executed upon it. In place of the positive expression of the difference, namely the increment, the negative expression of the increment, namely the difference, and thus $f(x+h)-f(x)$, therefore comes to the fore on the left-hand side. And this emphasis on the difference instead of the increment ('fluxion' in Newton) is foreshadowed at least in the *dy* of Leibnitzian notation as opposed to the Newtonian $\dot{y}$.

2) $f(x+h)-f(x) = 3x^2h + 3xh^2 + h^3$.

When both sides have been divided by h, we obtain

$$\frac{f(x+h)-f(x)}{h} = 3x^2 + 3xh + h^2 .$$

Thereby is formed on the left-hand side

$$\frac{f(x+h)-f(x)}{h} = \frac{(x+h)-f(x)}{x_1 - x}$$

which therefore *appears as a derived ratio of finite differences*, while with the mystics it was a completed ratio of increments given by the definitions of dx or $\dot{x}$ and dy or $\dot{y}$.

3) Now when in

$$\frac{f(x+h)-f(x)}{h} = \frac{f(x+h)-f(x)}{x_1 - x}$$

h is set $= 0$, or $x_1 = x$ so that $x_1 - x = 0$, this expression is transformed to $\frac{dy}{dx}$, while by means of this setting $h = 0$ the terms $3xh + h^2$ all become [zero] simultaneously, and this by means of a correct mathematical operation. They are thus now discarded without sleight of hand. One obtains:

4) $\frac{0}{0}$ or $\frac{dy}{dx} = 3x^2 = f'(x)$.

Just as with the mystics, this already existed as given, as soon as x became $x+h$, for $(x+h)^3$ in place of x^3 produces $x^3 + 3x^2h +$ etc., where $3x^2$ already appears in the second term of the series as the *coefficient* of h to the first power. The derivation is therefore essentially [the] same as in Leibnitz and Newton, but the ready-made derivative $3x^2$ is *separated* in a strictly algebraic manner from its other companions. It is no *development* but rather a *separation* of the $f'(x)$ — here $3x^2$ — from its factor h and from the neighbouring terms marching in closed ranks in the series. What has on the other hand really been developed is the left-hand, symbolic side, namely dx, dy, and their ratio, the symbolic differential coefficient $\frac{dy}{dx} = \frac{0}{0}$ (rather the inverse, $\frac{0}{0} = \frac{dy}{dx}$), which in turn once more generates certain metaphysical shudderings, although the symbol has been mathematically derived.

D'Alembert stripped the mystical veil from the differential calculus and took an enormous step forward. Although his *Traité des fluides* appeared in 1744 (see p.15*), the Leibnitzian method continued to prevail for years in France. It is hardly necessary to remark that Newton prevailed in England until the first decades of the 19th century. But here as in France earlier d'Alembert's foundation has been dominant until today, with some modifications.

3) Purely Algebraic Differential Calculus. Lagrange, '*Théorie des fonctions analytiques*' (1797 and 1813). Just as under *1)* and *2)*, the first starting point is the increased x; if

$$y \text{ or } f(x) = \text{etc.},$$

then it is y_1 or $f(x + dx)$ in the mystical method, y_1 or $f(x + h)$ $(= f(x + \Delta x))$ in the rational one. This binomial starting point immediately produces the binomial expansion on the other† side, for example:

$$x^{m} + mx^{m-1}h + \text{ etc.},$$

where the second term $mx^{m-1}h$ already yields ready-made the real differential coefficient sought, mx^{m-1}.

a) When $x + h$ replaces x in a given original function of x, $f(x + h)$ is related to the series expansion (*Entwicklungsreihe*) opposite it in exactly the same way that the *undeveloped general expression* in algebra, in particular the binomial, is related to its corresponding *series expansion*, as $(x + h)^3$, for example in

$$(x + h)^3 = x^3 + 3x^2h + \text{ etc.},$$

---

* See p.76

† i.e. right-hand — *Trans.*

is related to its equivalent series expansion $x^3 + 3x^2h +$ etc. With that step $f(x+h)$ enters into the very same algebraic relationship (only using variable quantities) which the general expression has toward its expansion throughout algebra, the relationship, for example, which $\frac{a}{a-x}$ in

$$\frac{a}{a-x} = 1 + \frac{x}{a} + \frac{x^2}{a^2} + \frac{x^3}{a^3} + \text{ etc.,}$$

has toward the series expansion $1 +$ etc., or which $\sin(x+h)$ in

$$\sin(x+h) = \sin x \cos h + \cos x \sin h$$

has toward the expansion standing opposite it.

D'Alembert merely algebraicised $(x + dx)$ or $(x + \dot{x})$ into $(x + h)$, and thus $f(x+h)$ from $y + dy$, $y + \dot{y}$ into $f(x+h)$. But Lagrange reduces the entire expression (*Gesamtausdruck*) to a purely algebraic character, since he places it, as a *general underdeveloped expression*, opposite the series expansion to be derived from it.

b) In the first method *1)*, as well as the rational one *2)*, the real coefficient sought is fabricated ready-made by means of the binomial theorem; it is found at once in the second term of the series expansion, the term which therefore is necessarily combined with h^1. All the rest of the differential process then, whether in *1)* or in *2)*, is a luxury. We therefore throw the needless ballast overboard. From the binomial expansion we know once and for all that the first real coefficient is the factor of h, the second that of h^2, and so on. The real differential coefficients are nothing other than those of the binomially developed series of *derived functions of the original function* in x (and the introduction of this category of *derived function* one of the most important). As for the separate differential forms, we know that Δx is transformed into dx, Δy into dy, and that the symbolic figure of $\frac{dy}{dx}$ represents the first derivative, the symbolic figure $\frac{d^2y}{dx^2}$ represents the second derivative, the coefficient of $\frac{1}{2}h^2$, etc. We may thus allow the symmetry of half of our purely algebraically obtained result to appear at the same time in these its differential equivalent quantities (*Differentialäquivalenten*)—a matter of nomenclature alone, all that remains from differential calculus proper. The whole problem is then resolved into finding (algebraic) methods 'of developing all kinds of functions of $x + h$ in integral ascending powers of h, which in many cases cannot be effected without great prolixity of operation'.[68]

Until this point there is nothing in Lagrange which could not be a direct result of d'Alembert's method (since this includes also the entire development of the mystics, only corrected).

c) While the development, therefore, of y_1 or $f(x + h) =$ etc. steps into the place of the differential calculus up to now ⟦and thereby, in fact, clarifies the mystery of the methods proceeding from

$$y + dy \text{ or } y + \dot{y},\ x + dx \text{ or } x + \dot{x}\ ,$$

namely that their real development rests on the application of the binomial theorem, while they represent from the very beginning the increased x_1 as $x + dx$, the increased y_1 as $y + dy$, and thus transform a monomial into a binomial⟧, the task now becomes, since we have in $f(x + h)$ a function without degree before us, the *general undeveloped expression* itself only, to derive algebraically from this undeveloped expression the general, and therefore valid for all power functions of x, *series expansion*.

Here Lagrange takes as his immediate starting point for the algebraicisation of the differential calculus the theorem of *Taylor outlived by Newton and the Newtonians*[69] which in fact is the most general, comprehensive theorem and at the same time operational formula of differential calculus, namely the series expansion, expressed in symbolic differential coefficients, of y_1 or $f(x + h)$, viz:

y_1 or $f(x + h)$

$$= y\,(\text{or } f(x)) + \frac{dy}{dx}h + \frac{d^2y}{dx^2}\frac{h^2}{[2]} + \frac{d^3y}{dx^3}\frac{h^3}{[2.3]} + \frac{d^4y}{dx^4}\frac{h^4}{[2.3.4]} + \text{etc.}$$

d) Investigation of Taylor's and MacLaurin's theorem to be added here.[70]

e) Lagrange's algebraic expansion of $f(x+h)$ into an equivalent series, which Taylor's $\frac{dy}{dx}$ etc. replaces, and it may only still be the symbolic differential expression of the algebraically derived functions of x. (This is to be continued from here on.[71])

III CONTINUATION OF EXTRACTS

c) Continuation of [*p.*] *25**

We have $x_1 - x = \Delta x$ from the beginning for the expression of the *difference* $x_1 - x$; the *difference exists here only in its form as a difference* (as, if y is dependent on x, $y_1 - y$ is written for the most part). Since we set $x_1 - x = \Delta x$, we already give the difference an expression different from itself. We express, if only in indeterminate form, the *value of this difference* as a quantity distinct from the difference itself. For example, $4-2$ is the pure expression of the difference between 4 and 2; but $4-2 = 2$ is the difference expressed in 2 (on the right-hand side): a) in positive form, so no longer as the difference; b) the subtraction is completed, the difference is calculated, and $4-2 = 2$ gives me $4 = 2+2$. The second 2 appears here in the positive *form of the increment* of the *original* 2. Therefore in *a form directly opposite to the difference form* (*einer der Differenzform entgegengesetzten Form*). Just as $a - b = c$, $a = b + c$, where c appears as the increment of b, so in $x_1 - x = \Delta x$, $x_1 = x + \Delta x$, where Δx enters immediately as the increment of x.

The simple original setting $x_1 - x = \Delta x =$ anything† therefore puts in place of the *difference form* another form, indeed that of a sum, $x_1 = x + \Delta x$, and at the same time simply expresses the difference $x_1 - x$ as the equivalent of the value of this difference, the quantity Δx.

It's just the same in $x_1 - x = \Delta x$, $x_1 - \Delta x = x$. We have the difference form again here on the left-hand side, but this time as the difference between the increased x_1 and its own increment,

* See p.84.

† In English in the original.

standing independent next to it. The difference between it and the increment of $x(=\Delta x)$ is a difference which now already expresses a defined, if also indeterminate, value of x.

If however one leaves the mystical differential calculus, where $x_1 - x$ enters immediately as $x_1 - x = dx$, and one first of all* corrects dx to Δx, then one begins from $x_1 - x = \Delta x$; thus from $x_1 = x + \Delta x$; but this in turn may then be turned round to $x + \Delta x = x_1$, so that the increase of x again attains the undefined form x_1, and as such enters directly into the calculus. This is the starting point of our applied algebraic method.

d) From this simple distinction of form there immediately results a fundamental difference in the treatment of the calculus which we demonstrate in detail (see the enclosed loose sheets)[72] in the analysis of d'Alembert's method. Here we have only to remark in general:

1) If the *difference* $x_1 - x$ (and thus $y_1 - y$) enters immediately as its opposite, *as the sum* $x_1 = x + \Delta x$ with its value therefore immediately in *the positive form of the increment* Δx, then, if x is replaced by $x + \Delta x$ everywhere in the *original function* in x, a binomial of definite degree is developed and the development of x_1 is resolved into an *application of the binomial theorem*. The binomial theorem is nothing but the general expression which results from a binomial of the first degree *multiplied* by itself m times. *Multiplication* therefore becomes the method of development of x_1 [or] $(x + \Delta x)$ if from the beginning we interpret the *difference* as *its opposite*, as *a sum*.

2) Since in the general form $x_1 = x + \Delta x$ the difference $x_1 - x$, in its positive form Δx, in the form, that is of the *increment*, is the *last or second* term of the expression, thus x becomes the first and Δx the second term of the original function in x when this is presented as a function in $x + \Delta x$. We know from the binomial theorem, however, that the second term only appears next to the first term as a factor raised to increasing powers, as a multiplier, so that the factor of the first

---

* Original *d'abord* — *Trans.*

expression in x (which is determined by the degree of the binomial) is $(\Delta x)^0 = 1$, the multiplier of the second term is $(\Delta x)^1$, that of the third is $(\Delta x)^2$, etc. The difference, in the positive form of the increment, therefore only comes in as a multiplier, and then for the first time, really (since $(\Delta x)^0 = 1$), as the multiplier of the second term of the expanded binomial $(x+\Delta x)^m$.

3) If on the other hand we consider the development of the function in x itself, the binomial theorem then gives us for this first term, here x, the series of its derived functions. For example, if we have $(x + h)^4$, where h is the known quantity in the binomial and x the unknown, we then have

$$x^4 + 4x^3h + \text{ etc.}$$

$4x^3$, which appears in the second term and has the factor h raised to the first power, is thus the first derived function of x, or, expressed algebraically: if we have $(x + h)^4$ as the *undeveloped expression of the binomial*, then the developed series gives us for the first increase of x^4 (for the increment) $4x^3$, which enters as the coefficient of h. If, however, x is a variable quantity and we have $f(x) = x^4$, then this by its very growth becomes $f(x + h)$, or, in the first form,

$$f(x+\Delta x) = (x+\Delta x)^4 = x^4 + 4x^3\Delta x + \text{ etc.}$$

x^4, which is provided for us in the usual algebraic binomial $(x + h)^4$ as the first term of the binom[ial expansion], now appears in the binomial expression of the variable, in $(x+\Delta x)^4$, as the reproduction of the original function in x before it increased and became $(x+\Delta x)$. It is clear from the very beginning by the nature of the binomial theorem that when $f(x) = x^4$ becomes $f(x + h) = (x + h)^4$, the first member of [the expansion of] $(x + h)^4$ is equal to x^4, that is, must be = the original function in x; $(x + h)^4$ must contain both the original function in x (here x^4) + the addition of all the terms which x^4 gains by becoming $(x + h)^4$, and thus the first term [of the expansion of] of the binomial $(x + h)^4$ [is the original function].

4) Furthermore: the second term of the binomial expansion, $4x^3h$, provides us immediately *ready-made* (*fix und fertig*) with the first derived function of x^4, namely $4x^3$. Thus this derivation has been obtained by the expansion of

$$f(x+\Delta x) = (x+\Delta x)^4;$$

obtained by means of the interpretation from the beginning of the difference $x_1 - x$ as its *opposite*, as the *sum* $x + \Delta x$.

It is thus the binomial expansion of $f(x + \Delta x)$, or y_1, which $f(x)$ has become by its increase, which gives us the first derivative, the coefficient of h (in the binomial series); and indeed right at the beginning of the binomial expansion, in its second term. The derivative is thus in no way obtained by differentiation but instead simply by the expansion of $f(x + h)$ or y_1 into a defined expression obtained by simple multiplication.

The crucial point (*Angelpunkt*) of this method is thus the development from the undefined expression y_1 or $f(x + h)$ to the defined binomial form, but using not at all the development of $x_1 - x$ and therefore as well of $y_1 - y$ or $f(x + h) - f(x)$ as differences.

5) The only difference equation which comes out in this method is the one which we obtain immediately:

$$f(x+\Delta x) = (x+\Delta x)^4 = x^4 + 4x^3\Delta x + 6x^2\Delta x^2 + 4x\Delta x^3 + \Delta x^4,$$

when we write:

$$x^4 + 4x^3\Delta x + 6x^2\Delta x^2 + 4x\Delta x^3 + \Delta x^4 - x^4,$$

putting the original function x^4, which forms the beginning of the series, back again behind, we now have before us the *increment* which the original function in x obtained through the use of the binomial expansion. Newton also writes in this way. And so we have the increment

$$4x^3\Delta x + 6x^2\Delta x^2 + 4x\Delta x^3 + \Delta x^4,$$

the increment of the original function, x^4. This way we use, on the other hand, *no difference of any kind*. The increment of y comes from the increment of x, if

$$y \text{ or } f(x) = x^4 .$$

So that Newton also writes immediately:

$$dy\text{, \textit{to him} } \dot{y} = 4x^3\dot{x} + \text{ etc.}$$

6) The entire remaining development now consists of the fact that we have to liberate the ready-made derivative $4x^3$ from its factor Δx and from its neighbouring terms, to prise it loose from its surroundings. So this is no method of development, but rather a *method of separation*.

e) The differentiation of $f(x)$ (as [a] general expression)

Let us note first of all (*d'abord*) that the concept of the 'derived function', for the successive real equivalents of the symbolic differential coefficients, which was completely unknown to the original discoverers of differential calculus and their first disciples, was in fact first introduced by Lagrange. To the former the dependent variable, y for example, appears only as a *function of* x, corresponding completely to the original algebraic meaning of *function*, first applied to the so-called indeterminate equations where there are more unknowns than equations, where therefore y, for example, takes on different values as different values are assumed for x. With Lagrange, however, the original function is the defined expression of x which is to be differentiated; so if y or $f(x) = x^4$, then x^4 is the original function, $4x^3$ is the first derivative, etc. In order to lessen the confusion, then, the dependent y or $f(x)$ is to be called the *function of* x in contrast to the original function in the Lagrangian sense, the *original function in* x, corresponding to the 'derived' functions *in* x.

In the algebraic method, where we first develop f^1, the preliminary derivative or [the ratio of] finite differences, and where we first develop from it the definitive derivative, f', we know from the very beginning: $f(x) = y$, so that a) $\Delta f(x) = \Delta y$, and therefore turned round $\Delta y = \Delta f(x)$. What is developed next is just $\Delta f(x)$, the value of the finite difference of $f(x)$.

We find

$$f^1(x) = \frac{\Delta y}{\Delta x}\ , \text{ so that } \frac{\Delta y}{\Delta x} = f^1(x)\ .$$

And so as well:

$$\Delta y = f^1(x)\Delta x\ ,$$

and since $\Delta y = \Delta f(x)$,

$$\Delta f(x) = f^1(x)\Delta x\ .$$

The next development of the differential expression, which finally yields

$$df(x) = f'(x)\,dx\ ,$$

is simply the differential expression of the previously developed finite difference.

In the usual method

$$dy \text{ or } df(x) = f'(x)\,dx$$

is not developed at all, rather instead, see above, the $f'(x)$ provided ready-made by the binomial $(x+\Delta x)$ or $(x + dx)$ is only *separated* from its factor and its neighbouring terms.

Taylor's Theorem, MacLaurin's Theorem and Lagrange's Theory of Derived Functions

1. FROM THE MANUSCRIPT 'TAYLOR'S THEOREM, MACLAURIN'S THEOREM, AND LAGRANGE'S THEORY OF DERIVED FUNCTIONS'[73]

I

Newton's discovery of the binomial (in his application, also of the polynomial) theorem revolutionised the whole of algebra, since it made possible for the first time a *general theory of equations*.

The binomial theorem, however — and this the mathematicians have definitely recognised, particularly since Lagrange — is also the primary basis (*Hauptbasis*) for differential calculus. Even a superficial glance shows that outside the circular functions, whose development comes from trigonometry, all differentials of monomials such as x^m, a^x, $\log x$, etc. can be developed from the binomial theorem alone.[74]

It is indeed the fashion of textbooks (*Lehrbuchsmode*) nowadays to prove both that the binomial theorem can be derived from Taylor's and MacLaurin's theorems and the converse.[75] Nonetheless nowhere — not even in Lagrange, whose theory of derived functions gave differential calculus a new foundation (*Basis*) — has the connection between the binomial theorem and these two theorems been established in all its original simplicity, and it is important here as everywhere, for science to strip away the veil of obscurity.

Taylor's theorem, historically prior to that of MacLaurin's, provides — under certain assumptions — for any function of x which increases by a positive or negative increment h,[76] therefore in general for $f(x \pm h)$, a series of symbolic expressions indicating by what series of differential operations $f(x \pm h)$ is to be developed. The subject at hand is thus the development of an arbitrary *function of x, as soon as it varies*.

MacLaurin on the other hand — also under certain assumptions — provides the general development of *any function of x itself*, also in a series of symbolic expressions which indicate how such functions, whose solution is often very difficult and complicated algebraically, can be found easily by means of differential calculus. The development of an arbitrary function of x, however, means nothing other than the *development of the constant functions combined with* [powers of] *the independent variable x*,[77] for the development of the variable itself should be identical to its variation, and thus to the object of Taylor's theorem.

Both theorems are grand generalisations in which the differential symbols themselves become the contents of the equation. In place of the real successive derived functions of x only the derivatives are represented, in the form of their symbolic equivalents, which indicate just so many strategies of operations to be performed, independently of the form of the function of $f(x+h)$. And so two formulae are obtained which with certain restrictions are applicable to all specific functions of x or $x+h$.

Taylor's Formula:

$$f(x+h) \text{ or } y_1$$

$$= y + \frac{dy}{dx}h + \frac{d^2y}{dx^2}\frac{h^2}{1.2} + \frac{d^3y}{dx^3}\frac{h^3}{1.2.3} + \frac{d^4y}{dx^4}\frac{h^4}{1.2.3.4} + \text{etc}.$$

MacLaurin's Formula:

$$f(x) \text{ or } y =$$

$$= (y) + \left(\frac{dy}{dx}\right)\frac{x}{1} + \left(\frac{d^2y}{dx^2}\right)\frac{x^2}{1.2} + \left(\frac{d^3y}{dx^3}\right)\frac{x^3}{1.2.3} + \left(\frac{d^4y}{dx^4}\right)\frac{x^4}{1.2.3.4}$$

$$+ \text{ etc}.$$

The mere appearance here shows what one might call, both historically and theoretically, the *arithmetic of differential calculus*, that is, the development of its fundamental operations is already assumed to be well-known and available. This should not be forgotten in the following, where I assume this acquaintance.

II

. . . .

MacLaurin's theorem may be treated as a *special case* of Taylor's theorem.

With Taylor we have

$$y = f(x),$$

$$y_1 = f(x+h) = f(x) \text{ or } y + \frac{dy}{dx}h + \frac{1}{2}\frac{d^2y}{dx^2}h^2 + \text{etc.}$$

$$+ \left[\frac{1}{1.2.3\ldots n}\right]\frac{d^ny}{dx^n}h^n + \text{etc.}$$

If we set $x = 0$ in $f(x+h)$ and on the right-hand side as well, in y or $f(x)$ and in its symbolic derived functions of the form $\frac{dy}{dx}, \frac{d^2y}{dx^2}$ etc., so that they consist simply of the development of the constant elements of x,[78] then:

$$f(h) = (y) + \left(\frac{dy}{dx}\right)h + \left(\frac{d^2y}{dx^2}\right)\frac{h^2}{1.2} + \left(\frac{d^3y}{dx^3}\right)\frac{h^3}{1.2.3} + \text{etc.}$$

$y_1 = f(x+h) = f(0+h)$ then becomes the same function of h which $y = f(x)$ is of x; since h goes into $f(h)$ just as x goes into $f(x)$ and (y) into $\left(\frac{dy}{dx}\right)$, all trace of x is wiped out.

We therefore can replace h with x on both sides and then obtain:

$$f(x) = (y) \text{ or } f(0) + \left(\frac{dy}{dx}\right)x + \left(\frac{d^2y}{dx^2}\right)\frac{x^2}{1.2} + \text{etc.}$$
$$+ \left(\frac{d^ny}{dx^n}\right)\frac{x^n}{1.2.3\ldots n} + \text{etc.}$$

Or as others have written it,

$$f(x) = f(0) + f'(0)x + f''(0)\frac{x^2}{1.2} + f'''(0)\frac{x^3}{1.2.3} + \text{etc.}$$

such as for example in the development of $f(x)$ or $(c + x)^m$:

$$(c + 0)^m = f(0) = c^m,$$
$$m(c + 0)^{m-1}x = mc^{m-1}x = f'(0)x \text{ etc.}$$

In the following, where we come to Lagrange, I will no longer consider MacLaurin's theorem as merely a special case of Taylor's. Let it only be noted here that it has its so-called 'failures'* just like Taylor's theorem. The failures all originate in the former in the irrational nature of the constant function, in the latter in that of the variable.[79]

It may now be asked:

Did not Newton merely give the result to the world, as he does, for example, in the most difficult cases in the *Arithmetica Universalis*, having already developed in complete silence Taylor's and MacLaurin's theorems for his private use from the binomial theorem, which he discovered? This may be answered with absolute certainty in the negative: he was not one to leave to his students the credit (*Aneignung*) for such a discovery. In fact he was still too absorbed in working out the differential operations themselves, operations which are already assumed to be given and well-known in Taylor and MacLaurin. Besides, Newton, as his first elementary formulae of calculus show, obviously arrived at them at first from mechanical points of departure, not those of pure analysis.

* In English in quotes in the original — *Trans.*

As for Taylor and MacLaurin on the other hand, they work and operate from the very beginning on the ground of differential calculus itself and thus had no reason (*Anlass*) to look for its simplest possible algebraic starting-point, all the less so since the quarrel between the Newtonians and Leibnitzians revolved about the defined, already completed forms of the calculus as a newly discovered, completely separate discipline of mathematics, as different from the usual algebra as Heaven is wide (*von der gewöhnlichen Algebra himmelweit verschiednen*).

The relationship of their respective *starting equations* to the binomial theorem was understood for itself, but no more than, for example, it is understood by itself in the differentiation of xy or $\frac{x}{y}$ that these are expressions obtained by means of ordinary algebra.

The real and therefore the simplest relation of the new with the old is discovered as soon as the new gains its final form, and one may say the differential calculus gained this relation through the theorems of Taylor and MacLaurin. Therefore the thought first occurred to Lagrange to return the differential calculus to a firm algebraic foundation (*auf strikt algebraische Basis*). Perhaps his forerunner in this was *John Landen*, an English mathematician from the middle of the 18th century, in his *Residual Analysis*. Indeed, I must look for this book in the [British] Museum before I can make a judgement on it.

III. Lagrange's Theory of Functions

Lagrange proceeds from the algebraic basis (*Begründung*) of Taylor's theorem, and thus from the most general formula of differential calculus.

It is only too noticeable with respect to Taylor's beginning equation:

$$y_1 \text{ or } f(x + h) = y \text{ or } f(x) + Ah + Bh^2 + Ch^3 + \text{etc.}$$

1) This series is in no way proved; $f(x + h)$ is no binomial of a *defined* degree; $f(x + h)$ is much more the undefined general expression of any function [of the variable] x which increases

by a positive or negative increment h; $f(x+h)$ therefore includes functions of x of any defined degree but at the same time excludes any defined degree to the series expansion itself. Taylor himself therefore puts '+ etc.' on the end of the series. However, that the series expansion which is valid for defined functions of x containing an increment — whether they are capable of representation now in a finite equation[80] or an infinite series — is no longer applicable to the undefined general $f(x)$ and therefore equally well to the undefined general $f(x_1)$ or $f(x+h)$, must first be *proved*.

2) The equation is translated into the language of differentials by virtue of the fact that it is twice differentiated, that is, y_1 once with respect to h as variable and x constant, but then again with respect to x as variable and h constant. In this manner two equations are produced whose first sides are identical while their second sides are different in form. In order, however, to be able to equate the undefined coefficients (which are all functions of x) of these two sides, it is also necessary to assume both that the individual coefficients A, B, etc., are *undefined, to be sure, but finite quantities*, and that their accompanying factors h increase in *whole and positive powers*.[81] If it is assumed — which is not the case — that Taylor had proved everything for $f(x+h)$ as long as the x in $f(x)$ remains *general*, then for that very reason it would not be valid at all as soon as the functions of x took on definite, particular values. This could be on the contrary irreconcilable with the treatment, by means of its series,

$$y_1 = y + \frac{dy}{dx}h + \frac{d^2y}{dx^2}h^2 + \text{etc.}$$

In one word, the conditions or assumptions which are involved in Taylor's unproven beginning equation are naturally found also in the theorem derived from it:

$$y_1 = y + \frac{dy}{dx}h + \frac{d^2y}{dx^2}h^2 + \text{etc.}$$

It is therefore inapplicable to certain functions of x which contradict any of the assumptions. Therefore the so-called *failures* of the theorem.

Lagrange provided an algebraic foundation for the beginning equation (*begründet die Ausgangsgleichung algebraisch*) and at the same time showed by means of the development itself which particular cases, due to their *general* character, that is, contradicting the general, undefined character of the function of x, are excluded.

H) 1) Lagrange's great service is not only to have provided a foundation in pure algebraic analysis for the Taylor theorem and differential calculus in general, but also and in particular to have introduced the concept of the derived function, which all of his successors have in fact used, more or less, although without mentioning it. But he was not satisfied with that. He provides the purely algebraic development of all possible functions of $(x + h)$ with increasing whole positive powers of h and then attributes to it the given name (*Taufname*) of the differential calculus. All the conveniences and condensations (Taylor's theorem, etc.) which differential calculus affords itself are thereby forfeited, and very often replaced by algebraic operations of much more far-reaching and complicated nature.

2) As far as pure analysis is concerned Lagrange in fact becomes free from all of what to him appears to be metaphysical transcendence in Newton's fluxions, Leibnitz's infinitesimals of different order, the limit value theorem of vanishing quantities, the replacement of $\frac{0}{0}\left(=\frac{dy}{dx}\right)$ as a symbol for the differential coefficient, etc. Still, this does not prevent him from constantly needing one or another of these 'metaphysical' representations himself in the application of his theories and curves etc.

2. FROM THE UNFINISHED MANUSCRIPT 'TAYLOR'S THEOREM'

If therefore in Taylor's theorem[82] 1) we adopt the idea from a *specific form* of the binomial theorem in which it is assumed that in $(x + h)^m$ m is a *whole positive power* and thus also that the factors appear as $h = h^0, h^1, h^2, h^3$, etc., that is, that h [is raised to a] whole, increasing, positive power, then 2) just as in the algebraic binomial theorem of the *general* form, the *derived functions of* x are defined and thereby *finite functions in* x. At this point, however, yet a third condition comes in. The derived functions of x can only be $= 0, = +\infty, = -\infty$, just as $h^{[k]}$ can only be $= h^{-1}$ or $h^{m/n}$ (for example $h^{1/2}$) when the variable x takes on *particular values*, $x = a$, for example.[83]

Summed up in general: *Taylor's theorem* is in general only applicable to the development of functions of x in which x becomes $= x + h$ or is increased from x to x_1 if 1) the independent variable x retains the *general, undefined form* x; 2) the original function in x itself is capable of development by means of differentiation into a series of *defined* and thereby *finite, derived functions in* x, with corresponding factors of h with increasing, positive and integral exponents, so with h^1, h^2, h^3 etc.

All these conditions, however, are only another expression for the fact that this theorem *is* the binomial theorem with *whole and positive exponents*, translated into differential language.

Where these conditions are not fulfilled, where *Taylor's theorem is not applicable*, that is, there enter what are called in differential calculus the '*failures*'* of this theorem.

---

* In English in original — *Ed.*

The biggest *failure* of Taylor's theorem, however, does not consist of these particular failures of application but rather *the general failure*, that

$$y = f(x) \quad [\text{and}] \quad y_1 = f(x+h)\,,$$

which are only symbolic expressions of a binomial of some sort of degree,[84] are transformed into expressions where $f(x)$ is a function of x which includes all degrees and thereby has *no degree* itself, so that $y_1 = f(x+h)$ equally well includes all degrees and is itself of no degree, and even more that it becomes the *undeveloped general expression* of any function of the variable x, as soon as it increases. The series development into which the ungraded $f(x+h)$ is expanded, namely $y + Ah + Bh^2 + Ch^3 +$ etc., therefore also includes all degrees without itself having any degree.

This leap from *ordinary algebra*, and besides *by means of ordinary algebra*, into the *algebra of variables* is assumed as *un fait accompli*, it is not proved and is prima facie *in contradiction to all the laws* of conventional algebra, where $y = f(x)$, $y_1 = f(x+h)$ could never have this meaning.

In other words, the starting equation

$$y_1 \text{ or } f(x+h) = y \text{ or } f(x) + Ah + Bh^2 + \\ + Ch^3 + Dh^4 + Eh^5 + \text{ etc.}$$

is not only *not proved* but indeed knowingly or unknowingly assumes a substitution of *variables* for *constants*, which flies in the face of all the laws of algebra — for algebra, and thus the algebraic binomial, only admits of constants, indeed only two sorts of constants, *known* and *unknown*. The derivation of this equation from algebra therefore appears to rest on a deception.

Yet now if in fact Taylor's theorem — whose failures in application hardly come into consideration, since as a matter of fact they are restricted to functions of x with which differentiation gives no result[85] and are thus in general inaccessible to treatment by the differential calculus — has proved to be in practice the most comprehensive, most general and most successful *operational formula* (*Operationsformel*) of all differential calculus; then this is only the crowning of the edifice of the Newtonian school, to which he belonged, and of the Newton-Leibnitz period of development of differential calculus in general, which from the very beginning drew correct results from false premises.

The algebraic proof of Taylor's theorem has now been given by *Lagrange*, and it in general provides the foundation (*Basis*) of *his* algebraic method of differential calculus. On the subject itself I will go into greater detail in the eventual historical part of this manuscript.[86]

As a *lusus historiae* [an aside in the story] let it be noted here that Lagrange in no way goes back to the unknown foundation for Taylor — to the binomial theorem, the binomial theorem in the most elementary form, too, where it consists of only two quantities, $(x + a)$ or here, $(x + h)$, and has a positive exponent.

Much less does he go back further and ask himself, why the binomial theorem of Newton, translated into differential form and at the same time freed of its algebraic conditions by means of a powerful blow (*Gewaltstreich*), appears as the comprehensive, overall operational formula of the calculus he founded? The answer was simple: because from the very beginning Newton sets $x_1 - x = dx$, so that $x_1 = x + dx$. The development of the *difference* is thus at once transformed into the development of a *sum* in the binomial $(x + dx)$ — whence we disregard completely that it had to have been set $x_1 - x = \Delta x$ or h (so that $x_1 = x + \Delta x$ or $= x + h$). Taylor only developed this fundamental basis to its most general and comprehensive form, which only became possible once all the fundamental operations of differential calculus had been discovered; for what sense had his $\frac{dy}{dx}$, $\frac{d^2y}{dx^2}$, etc. unless one could already develop the corresponding $\frac{dy}{dx}$, $\frac{d^2y}{dx^2}$, etc. for all essential functions in x?

Lagrange, conversely, bases himself directly on Taylor's theorem (*schliesst sich direkt an Taylor's Theorem an*), from a standpoint, naturally, where on the one hand the successors of the Newton-Leibnitz epoch already provide him with the corrected version of $x_1 - x = dx$, so that as well $y_1 - y = f(x+h) - f(x)$, while on the other hand he produced, right in the algebraicisation of Taylor's formula, his own theory of the *derived function*. [In just such a manner Fichte followed Kant, Schelling Fichte, and Hegel Schelling, and neither Fichte nor Schelling nor Hegel investigated the general foundation of Kant, of idealism in general: for otherwise they would not have been able to develop it further].

Appendices to the Manuscript 'On the History of the Differential Calculus' and Analysis of D'Alembert's Method

ON THE AMBIGUITY OF THE TERMS 'LIMIT' AND 'LIMIT VALUE'[87]

I) x^3;

a) $(x+h)^3 = x^3 + 3hx^2 + 3h^2x + h^3$;

b) $(x+h)^3 - x^3 = 3x^2h + 3xh^2 + h^3$;

c) $\frac{(x+h)^3 - x^3}{h} = 3x^2 + 3xh + h^2$.

If h becomes $= 0$, then

$$\frac{(x+0)^3 - x^3}{0} \text{ or } \frac{x^3 - x^3}{0} = \frac{0}{0} \text{ or } \frac{dy}{dx}$$

and the right-hand side $= 3x^2$, so that

$$\frac{dy}{dx} = 3x^2 .$$

$$y = x^3; \qquad y_1 = x_1^3 ;$$

$$y_1 - y = x_1^3 - x^3 = (x_1 - x)(x_1^2 + x_1x + x^2) ;$$

$$\frac{y_1 - y}{x_1 - x} \text{ or } \frac{dy}{dx} = x^2 + xx + x^2 ;$$

$$\frac{dy}{dx} = 3x^2 .$$

II) Let us set $x_1 - x = h$. Then:

1) $(x_1 - x)(x_1^2 + x_1x + x^2) = h(x_1^2 + x_1x + x^2)$;

2) so that:

$$\frac{y_1 - y}{h} = x_1^2 + x_1x + x^2 \ .$$

In 1) the coefficient of h is *not the completed derivative*, like f' above, but rather f^1; the division of both sides by h, therefore, also leads not to $\frac{dy}{dx}$, but rather

$$\frac{\Delta y}{h} \quad \text{or} \quad \frac{\Delta y}{\Delta x} = x_1^2 + x_1x + x^2$$

etc., etc.

If we begin on the other side in I c), namely in

$$\frac{f(x+h) - f(x)}{h} \quad \text{or} \quad \frac{y_1 - y}{h} = 3x^2 + 3xh + h^2 \ ,$$

from the assumption that the more the value of h decreases on the right-hand side, so much the more does the value of the terms $3xh + h^2$ decrease,[88] so that the value as well of the entire right-hand side $3x^2 + 3xh + h^2$ more and more closely approaches the value $3x^2$, we then must set down, however: 'yet without being able to coincide with it'.

$3x^2$ thus becomes a value which the series constantly approaches, without ever reaching it, and thus, even more, without ever being able to exceed it. In this sense $3x^2$ becomes the *limit value*[89] of the series $3x^2 + 3xh + h^2$.

On the other side the quantity $\frac{y_1 - y}{h}$ $\left(\text{or } \frac{y_1 - y}{x_1 - x}\right)$ also decreases all the more, the more its denominator h decreases.[90] Since, however, $\frac{y_1 - y}{h}$ is the equivalent of $3x^2 + 3xh + h^2$ the limit value of the series is also the ratio's own *limit* value in the same sense that it is the limit value of the equivalent series.

However, as soon as we set $h = 0$, the terms on the right-hand side vanish, making $3x^2$ the limit of its value; now $3x^2$ is the first derivative of x^3 and so $= f'(x)$. As $f'(x)$ it indicates that an $f''(x)$ is also derivable from it (in the given case it $= 6x$) etc., and thus that the increment $f'(x)$ or $3x^2$ is not = the sum of the increments which can be developed from $f(x) = x^3$. Were $f(x)$ itself an infinite series, so naturally the series of increments which can be developed from it would be infinite as well. In this sense, however, the developed series of increments becomes, as soon as I break it off, the *limit value* of the development, where *limit value* here is in the usual algebraic or arithmetic meaning, just as the developed part of an endless decimal fraction becomes the *limit* of its possible development, a limit which is satisfactory on practical or theoretical grounds. This has absolutely nothing in common with the limit value in the first sense.

Here in the second sense the limit value may be arbitrarily *increased*, while there it may be only *decreased*. Furthermore

$$\frac{y_1 - y}{h} = \frac{y_1 - y}{x_1 - x},$$

so long as h is only decreased, only approaches the expression $\frac{0}{0}$; this is a limit which it may never attain and still less exceed, and thus far $\frac{0}{0}$ may be considered its *limit value*.[91]

As soon, however, as $\frac{y_1 - y}{h}$ is transformed to $\frac{0}{0} = \frac{dy}{dx}$, the latter has ceased to be the limit value of $\frac{y_1 - y}{h}$, since the latter has itself disappeared into its limit.[92] With respect to its earlier form, $\frac{y_1 - y}{h}$ or $\frac{y_1 - y}{x_1 - x}$, we may only say that $\frac{0}{0}$ is its absolute minimal expression which, treated in isolation, is no expression of value (*Wertausdruck*); but $\frac{0}{0}$ $\left(\text{or } \frac{dy}{dx}\right)$ now has $3x^2$ opposite it as its real equivalent, that is $f'(x)$.

And so in the equation

$$\frac{0}{0}\left(\text{or }\frac{dy}{dx}\right)=f'(x)$$

neither of the two sides is the limit value of the other. They do not have a *limit relationship* (*Grenzverhältnis*) to one another, but rather a *relationship of equivalence* (*Äquivalentverhältnis*). If I have $\frac{6}{3} = 2$ then neither is 2 the limit of $\frac{6}{3}$ nor is $\frac{6}{3}$ the limit of 2. This simply comes from the well-worn tautology that the value of a quantity = the limit of its value.

The concept of the limit value may therefore be interpreted wrongly, and is constantly interpreted wrongly (*missdeutet*). It is applied in differential equations[93] as a means of preparing the way for setting $x_1 - x$ or $h = 0$ and of bringing the latter closer to its presentation: — a childishness which has its origin in the first mystical and mystifying methods of calculus.

In the application of differential equations to curves, etc., it really serves to make things more apparent geometrically.

COMPARISON OF D'ALEMBERT'S METHOD TO THE ALGEBRAIC METHOD

Let us compare d'Alembert's method to the algebraic one.[94]

I) $f(x)$ or $y = x^3$;

a) $f(x+h)$ or $y_1 = (x+h)^3 = x^3 + 3x^2h + 3xh^2 + h^3$;

b) $f(x+h) - f(x)$ or $y_1 - y = 3x^2h + 3xh^2 + h^3$;

c) $\dfrac{f(x+h) - f(x)}{h}$ or $\dfrac{y_1 - y}{h} = 3x^2 + 3xh + h^2$;

if $h = 0$:

d) $\dfrac{0}{0}$ or $\dfrac{dy}{dx} = 3x^2 = f'(x)$.

II) $f(x)$ or $y = x^3$;

a) $f(x_1)$ or $y_1 = x_1^3$;

b) $f(x_1) - f(x)$ or $y_1 - y = x_1^3 - x^3$

$$= (x_1 - x)(x_1^2 + x_1x + x^2);$$

c) $\dfrac{f(x_1) - f(x)}{x_1 - x}$ or $\dfrac{y_1 - y}{x_1 - x} = x_1^2 + x_1x + x^2$.

If x_1 becomes $= x$, then $x_1 - x = 0$, hence:

d) $\dfrac{0}{0}$ or $\dfrac{dy}{dx} = (x^2 + xx + x^2) = 3x^2$.

It is the same in both so far: if the independent variable x increases, so does the dependent [variable] y. Everything depends on how the increase of x is expressed. If x becomes x_1, then $x_1 - x = \Delta x = h$ (an undefined, infinitely contractible but always finite difference).[95]

If Δx or h is the increment by which x has increased, then: a) $x_1 = x + \Delta x$, but also in reverse b) $x + \Delta x$ or $x + h = x_1$.

The differential calculus begins historically with a); with the fact, that is, that the difference Δx or the increment h (one expresses the same thing as the other: the first negatively as the difference Δx, the second positively as the increment h) *exists independently* next to the quantity x, whose increment it is and thus which it expresses as *increased*, but increased by h. It thereby achieves the advantage from the very beginning, that the original function of the variables corresponding to this general expression, as soon as it increases, is expressed in a binomial of a defined degree, and therefore from the very beginning the binomial theorem is applicable to it. Already, in fact, we have a binomial on the general, the left-hand, side, namely $x + \Delta x$ [, such that $f(x + \Delta x)$] or $y_1 =$ etc.

The mystical differential calculus immediately transforms: $(x + \Delta x)$ into $(x + dx)$ or according to Newton, $x + \dot{x}$.[96] Thereby we have also immediately obtained on the right-hand, the algebraic, side a binomial in $x + dx$ or $x + \dot{x}$ which may be treated as an ordinary binomial. The transformation from Δx to dx or $\dot{x}$ is assumed *a priori* rather than rejected on mathematical grounds, so that later the mystical suppression of terms of the developed binomial becomes possible.

D'Alembert begins with $(x + dx)$ but corrects the expression to $(x + \Delta x)$, alias $(x + h)$; a development now becomes necessary in which Δx or h is transformed into dx, but all of that development really proceeds (*das ist auch alle Entwicklung, die wirklich vorgeht*).

Whether it begin falsely with $(x + dx)$ or correctly with $(x + h)$, this undefined binomial placed in the given algebraic function of x transforms into a binomial of a defined degree — such as $(x + h)^3$ now appears in Ia) instead of x^3 — and even into a binomial in which in the first case dx, in the other case h appears as its last term, and also in the expansion as well as merely a factor to which the functions derived from the binomial are externally attached (*behaftet*).

Therefore we find *right in* Ia) the complete *first derivative* of x^3, namely $3x^2$, as the coefficient in the second term of the series, attached to h. $3x^2 = f'(x)$ remains unchanged from now on. It is itself derived by means of no sort of process of differentiation at all but rather provided from the very beginning by means of the binomial theorem, indeed because from the very beginning we have represented the increased x as a binomial,

$$x + \Delta x = x + h ,$$

as x increased by h. The entire problem now consists of uncoupling not the embryonic but the ready-made $f'(x)$ from its factor h and from its other neighbouring terms.

In IIa) in contrast, the increased x_1 enters the algebraic function in exactly the same form as x originally entered it; x^3 becomes x_1^3. The derivative $f'(x)$ can only be obtained at the end by means of two successive differentiations, and those of quite distinct character indeed.

In equation Ib) the difference $f(x + h) - f(x)$ or $y_1 - y$ now prepares the arrival of the symbolic differential coefficient; in real terms, however, all that changes is that it moves out of second rank into the first rank of the series and therefore makes possible its liberation from h.

In IIb) we obtain the expression of differences on both sides; it has been so developed on the algebraic side that $(x_1 - x)$ appears as a factor beside a derived function in x and x_1 which was obtained by means of the division of $x_1^3 - x^3$ by $x_1 - x$. Only the existence of the difference $x_1^3 - x^3$ made possible its separation into two factors. Since

$$x_1 - x = h ,$$

the two factors into which $x_1^3 - x^3$ is resolved may also be written $h(x_1^2 + x_1x + x^2)$. This points up a new difference with Ib). h itself as the *factor of the preliminary derivative* is only derived by means of the expansion of the difference $x_1^3 - x^3$ into the product of two factors, while h as the factor of the 'derivative', exists just like the latter in Ia), already complete before any difference has been developed at all. That the undefined increase from x to x_1 takes the separated form of the factor h next to x, is assumed from the very beginning in I), but proved (since $x_1 - x = h$) by means of the derivation in II). Indeed, on the one hand h is undefined in I) while on the other hand it is already fairly well defined, since the undefined increase of x already appears as a *separate* quantity *by which* x has increased, and thus as such it enters next to it.

In Ic), $f'(x)$ is now freed of its factor h; we thus obtain on the left-hand side $\frac{y_1 - y}{h}$ or $\frac{f(x+h) - f(x)}{h}$, thus a still finite expression of the differential coefficient. On the other side, however, we have reached the point where, when we set $h = 0$ in $\frac{f(x+h) - f(x)}{h}$, and this transforms into $\frac{0}{0} = \frac{dy}{dx}$, we obtain on one side in Id) the symbolic differential coefficient and on the other $f'(x)$, which appeared complete already in Ia) but now has been freed of its neighbouring terms and stands alone on the right-hand side.

Positive development only proceeds on the left-hand side, since here the symbolic differential coefficient is produced. On the right-hand side the development consists only of freeing $f'(x) = 3x^2$, already found in Ia) by means of the binomial, from its original impediment. The transformation of h into 0 or $x_1 - x = 0$ has only this negative meaning on the right-hand side.

In IIc), by contrast, a *preliminary derivative* is only obtained by dividing both sides by $x_1 - x$ (= h).

Finally, in IId) the *definitive derivative* is obtained by the positive setting of $x_1 = x$. This $x_1 = x$ means, however, setting at the same time $x_1 - x = 0$, and therefore transforms the finite ratio $\frac{y_1 - y}{x_1 - x}$ on the left-hand side to $\frac{0}{0}$ or $\frac{dy}{dx}$.

In I) the 'derivative' is no more found by setting $x_1 - x = 0$ or $h = 0$ than it is in the mystical differential method. In both cases the neighbouring terms of the $f'(x)$ which appeared complete from the very beginning have been tossed aside, now in a mathematically correct manner, there by means of a coup d'etat.

ANALYSIS OF D'ALEMBERT'S METHOD BY MEANS OF YET ANOTHER EXAMPLE[97]

Let us now develop according to d'Alembert's method:

a) $f(u)$[98] or $y = 3u^2$;

b) $f(x)$ or $u = x^3 + ax^2$.

$$y = 3u^2 \,, \tag{1}$$

$$f(u) = 3u^2 \,. \tag{1a}$$

$$f(u+h) = 3(u+h)^2 \,,$$

$$\begin{aligned} f(u+h) - f(u) &= 3(u+h)^2 - 3u^2 \\ &= 3u^2 + 6uh + 3h^2 - 3u^2 = 6uh + 3h^2 \quad (2) \end{aligned}$$

(*here is the derived function, already complete in the coefficient of h by means of the binomial theorem*),

$$\frac{f(u+h) - f(u)}{h} = 6u + 3h \,.$$

$f'(u) = 6u$, already given complete in (2), is freed of its factor h by means of division.

$$\frac{f(u+0) - f(u)}{0} = 6u \,,$$

$$\frac{y_1 - y}{u_1 - u}, \text{ alias } \frac{0}{0} = \frac{dy}{du} = 6u \,.$$

Substituting in here the value of u from equation b) gives

$$\frac{dy}{du} = 6(x^3 + ax^2) .$$

Since y in a) is differentiated with respect to u, thus

$$(u_1 - u) = h \quad \text{or} \quad h = (u_1 - u) ,$$

since u is the independent variable.

And so:

$$\frac{dy}{du} = 6(x^3 + ax^2) .$$

(This is obtained from $f(u)$ or $y = 3u^2$.)

[We now develop b) in the same manner, so that]

b) $f(x)$ or $u = x^3 + ax^2$,

$$f(x + h) = (x + h)^3 + a(x + h)^2 ,$$

$$f(x + h) - f(x) = (x + h)^3 + a(x + h)^2 - x^3 - ax^2$$

$$= \begin{matrix} x^3 + 3x^2h + 3xh^2 + h^3 \\ + ax^2 + 2axh + ah^2 \end{matrix} \left\{ \begin{matrix} - x^3 \\ - ax^2 \end{matrix} \right.$$

$$= (3x^2 + 2ax)h + (3x + a)h^2 + h^3 ,$$

$$\frac{f(x + h) - f(x)}{h} = 3x^2 + 2ax + (3x + a)h + h^2 .$$

If we now set $h = 0$, on the second side:

$$\frac{0}{0} \quad \text{or} \quad \frac{du}{dx} = 3x^2 + 2ax .$$

The derived function is already contained complete, however, in

$$f(x + h) = (x + h)^3 + a(x + h)^2 ,$$

since this produces

$$x^3 + 3x^2h + 3xh^2 + h^3 + ax^2 + 2axh + ah^2 \ .$$

Thus

$$x^3 + ax^2 + (3x^2 + 2ax)h + (3x + a)h^2 + h^3 \ .$$

It already appears complete as the coefficient of h. This derivative is therefore not obtained by means of differentiation, but rather by means of an increase from $f(x)$ to $f(x+h)$ and thus from $x^3 + ax^2$ to $(x+h)^3 + a(x+h)^2$. It is obtained simply by virtue of the fact that when x becomes $x+h$ we obtain a binomial in $x+h$ of defined degree on the second side, a binomial whose second term, multiplied (*behaftetes*) by h, contains the derived function of u, $f'(u)$, ready-made (*fix und fertig*).

The rest of the procedures serve only to liberate the $f'(x)$ thus given from the very beginning from its own coefficient h and from all other terms.

The equation

$$\frac{f(x+h) - f(x)}{h} = \text{etc.}$$

provides two things: first, it makes it possible to obtain the numerator on the first side as the difference of $f(x)$, presently $= \Delta f(x)$; on the second side, however, it provides the algebraic opportunity to extract the original function given in x, $x^3 + ax^2$, from the product of $(x+h)^3 + a(x+h)^2$ etc.

So we continue. We have obtained for a):

$$\frac{dy}{du} = 6(x^3 + ax^2) \ ,$$

and for b):

$$\frac{du}{dx} = 3x^2 + 2ax \ .$$

We multiply $\frac{dy}{du}$ by $\frac{du}{dx}$, so that

$$\frac{dy}{du} \cdot \frac{du}{dx} = \frac{dy}{dx},$$

which was to be found. Let us substitute in here the values found for $\frac{dy}{du}$ and $\frac{du}{dx}$; so that

$$\frac{dy}{dx} = 6(x^3 + ax^2)(3x^2 + 2ax)$$

and therefore, generally expressed, if we have:

$$y = f(u); \quad \frac{dy}{du} = \frac{df(u)}{du}, \quad u = f(x); \quad \frac{du}{dx} = \frac{df(x)}{dx},$$

hence

$$\frac{dy}{du} \cdot \frac{du}{dx} \text{ or } \frac{dy}{dx} = \frac{df(u)}{du} \cdot \frac{df(x)}{dx}.$$

If we now substitute $h = u_1 - u$ into equation a) and $h = x_1 - x$ into equation b), things are so arranged that:

$$y \text{ or } f(u) = 3u^2,$$

$$\begin{aligned} f(u + (u_1 - u)) &= 3(u + (u_1 - u))^2 \\ &= 3u^2 + 6u(u_1 - u) + 3(u_1 - u)^2, \end{aligned}$$

$$\begin{aligned} f(u + (u_1 - u)) - f(u) &= 3u^2 + 6u(u_1 - u) \\ &\quad + 3(u_1 - u)(u_1 - u) - 3u^2, \end{aligned}$$

hence:

$$f(u + (u_1 - u)) - f(u) = 6u(u_1 - u) + 3(u_1 - u)^2,$$

$$\frac{f(u+(u_1-u))-f(u)}{u_1-u} = 6u + 3(u_1-u) .$$

Hence [if] $u_1 - u$ in the first term $= 0$, then

$$\frac{dy}{du} = 6u + 0 = 6u .$$

This shows that when $f(u)$ from the very beginning becomes $f(u + (u_1 - u))$, then its increment appears as the positive second term of a defined binomial on the second side, and this second term, which is multiplied by $(u_1 - u)$ or h by the binomial theorem, immediately becomes the coefficient to be found. If the second term is polynomial, as it is in

$$x^3 + ax^2 \text{, which becomes } (x+h)^3 + a(x+h)^2 ,$$

or

$$(x + (x_1 - x))^3 + a(x + (x_1 - x))^2 ,$$

then we have only to sum the terms multiplied by $x_1 - x$ to the first power, alias h to the first power, as the coefficient of h or $x_1 - x$; and we have again the complete coefficient.

This result shows:

1) that when in d'Alembert's development $x_1 - x = h$ is put in reverse $h = x_1 - x$, thereby absolutely nothing is changed in the method itself, rather the method simply brings out more clearly how to obtain the binomial by means of $f(x+h)$ or $f(x + (x_1 - x))$ for the algebraic expression on the other side in place of the original function, in place of $3u^2$ for example in the given case.

The second term which one finds in that manner attached to h or $(x_1 - x)$ is the complete first derived function. The problem now consists of freeing it of h or $x_1 - x$, which is easily done. There the derived function is complete; it is therefore not found by setting $x_1 - x = 0$, but rather freed of its factor

$(x_1 - x)$ and accessories. Just as it is found by simple multiplication (the binomial development) as the second term [with] $x_1 - x$, so it is finally freed of the latter by means of division of both sides by $x_1 - x$.

The crucial procedure (*Mittelprozedur*), however, consists of the development of the equation

$$f(x+h) - f(x) \quad \text{or} \quad f(x + (x_1 - x)) - f(x) = [\ldots] \, .$$

The equation has the sole purpose (*Zweck*) here of making the original function vanish on the second side, since the development [of] $f(x+h)$ necessarily contains $f(x)$ together with its increment developed by means of the binomial. This [$f(x)$] is thus extracted from the second side.

Therefore what happens, for example, in

$$(x+h)^3 + a(x+h)^2 - x^3 - ax^2 \, ,$$

is, that the first terms x^3 and ax^2 are extracted from the binomial $(x+h)^3 + a(x+h)^2$; we thus obtain, multiplied by h or $(x_1 - x)$, the already complete derived function as the first term of the equation.

The first differentiation on the second side is nothing but the simple subtraction of the original function from its increased expression, which thus gives us the increment by which it has increased and whose first term, multiplied by h, is already the complete derived function. The other terms can only contain h^2 etc. or $(x_1 - x)^2$ etc. as coefficients; they are reduced by one power with the first division of both sides by $x_1 - x$, while the first term emerges without any h.

2) The difference from the method of $f(x_1) - f(x) =$ etc. lies in the fact that, when we have for example

$$f(x) \quad \text{or} \quad u = x^3 + ax^2 \, ,$$

$$f(x_1) \quad \text{or} \; u_1 = x_1^3 + ax_1^2 \, ,$$

the first increment (*Anwachs*) of the variable x by no means provides us with $f'(x)$ ready-made from the very beginning.

$$f(x_1) - f(x) \quad \text{or} \quad u_1 - u = x_1^3 + ax_1^2 - (x^3 + ax^2) \ .$$

Here by no means is it a matter of extracting the original function again, since $x_1^3 + ax_1^2$ does not contain x^3 and ax^2 in any form. On the contrary, this first difference equation provides us with an opportunity for development (*Entwicklungsmoment*), namely the transformation of both of the two original terms into differences of [powers of] x_1 and x.

Namely,

$$= (x_1^3 - x^3) + a(x_1^2 - x^2) \ .$$

It is now clear that when we again resolve both of these two terms into factors of $x_1 - x$, we obtain functions in x_1 and x as coefficients of $x_1 - x$, namely:

$$f(x_1) - f(x) \quad \text{or} \quad u_1 - u = (x_1 - x)\,(x^2 + x_1 x + x^2)$$
$$+ \, a(x_1 - x)\,(x_1 + x) \ .$$

We divide this by $x_1 - x$, and the left-hand side as well, so that:

$$\frac{f(x_1) - f(x)}{x_1 - x} \quad \text{or} \quad \frac{u_1 - u}{x_1 - x} = (x_1^2 + x_1 x + x^2) + a(x_1 + x) \ .$$

By means of this division we have obtained the preliminary derivative. Each of its parts contains terms in x_1.

Thus we can finally obtain the first derived function in x only when we set $x_1 = x$, so that $x_1 - x = 0$, and then

$$x_1^2 = x^2 \ , \qquad x_1 x = x^2 \ ,$$

and thus:

$$(x_1^2 + x_1 x + x^2) = 3x^2 \quad \text{and} \quad x_1 + x = x + x = 2x \ ;$$

so that:

$$a(2x) = 2xa \ .$$

The result on the other [side]

$$\frac{df(x)}{dx} = \frac{du}{dx} = \frac{0}{0}.$$

Thus the derived function is here only obtained by setting $x_1 = x$, so that $x_1 - x = 0$. $x_1 = x$ provides the final positive result in the real function of x.

But $x_1 = x$ also leads to $x_1 - x = 0$ and therefore at the same time, beside this positive result, to the symbolic $\frac{0}{0}$ or $\frac{dy}{dx}$ on the other side.

We could have said from the very beginning: we have to obtain a *derivative* in x_1 and x in the end. This can only be transformed into the derivative in x when x_1 is set $= x$; but setting $x_1 = x$ is the same as setting $x_1 - x = 0$, which nullification is positively expressed by the formula $x_1 = x$ which is necessary for the transformation of the derivative to a function of x, while its negative form, $x_1 - x = 0$, must provide us with the symbol.

3) Even if this treatment of x, where an increment ($x_1 - x = \Delta x$, for example, or h) is not independently introduced next to it, was already well-known, something which is very probable and of which I shall convince myself by consulting J[ohn] Landen at the [British] Museum, still its essential difference cannot have been grasped.

What distinguishes this method from Lagrange, however, is that it really differentiates, so that the differential expression also originates on the symbolic side, while with him the derivation does not represent the differentiation algebraically, but instead derives the functions algebraically directly from the binomial and simply accepts their differential form 'by symmetry', since it is known from differential calculus that the first derived function $= \frac{dy}{dx}$, the second $= \frac{d^2y}{dx^2}$, etc.

Appendices by the editors of the Russian edition

APPENDIX I

Concerning the Concept of 'Limit' in the Sources consulted by Marx

In order to give the reader accustomed to the contemporary use in mathematics of the term 'limit' a correct understanding of Marx's critical remarks concerning this concept and of Marx's interpretation of it, we give first of all the definition of 'limit' (and clarifying examples) and the ways of using the word 'limit' contained in the courses of Hind and Boucharlat which Marx possessed and studied critically.

Hind's course-book follows d'Alembert, which is to say that the derivative was defined in it by means of the concept of limit. The introductory chapter of the book was therefore entitled 'The method of limits'. However, neither in this chapter nor in the rest of the textbook was there a definition of 'limit'. There were only definitions of the 'limits' of a variable in the restricted sense of the exact upper or lower bounds to the multiplicity of its value. (This multiplicity might include, in particular, an 'infinitely large' value of the variable, designated by the symbol ∞. But there were no precisely defined correct operations with this symbol: there was no concept of absolute value, no $+\infty$ and $-\infty$; it was considered simply self-evident, that for any $\alpha \geqslant 0$, $\infty + \alpha = \infty$, that for any finite a (that is, distinct from 0, as well as from ∞) $a . \infty = \infty$ and $\frac{a}{\infty} = 0$.)

This concept of the limit of a function — a concept which of course can only be surmised from the examples — was introduced in the introductory chapter, implicitly, by means, as might be anticipated, of identifying this limit (at the point coinciding with the exact upper or lower bounds of the given multiplicity of the values of the argument) with one of the 'limits' (with the exact upper or with the exact lower bound) of the corresponding multiplicity of values of the function. Since only monotonic or piecewise monotonic functions are examined in this book, such a 'limit' appears in practice to be with the (one-sided) limit in the more usual sense of the word, in which Hind actually uses the concept of limit in all the remaining parts of the book. It turned out, however, that the introduction of this concept, which was supposed to 'improve' the method of infinitely small quantities, did not consciously attain that goal and was generally unwarranted.

Actually, Hind might have replaced the evaluation of the one-sided limit of a piecewise monotonic function $f(x)$, defined on the interval (a, b) by the solution of the following two problems as x moves to $+a$:

1. To find a certain number α such that for $a < x < \alpha$ the function is monotonic (in the broad sense, i.e., non-decreasing or non-increasing; for demonstration we will assume the function is here monotonically non-decreasing);

2. To evaluate the point at the (by our assumption lower) boundary of the possible values of the function on the interval (a, α), that is, for $a < x < \alpha$. Clearly, this will be the desired $\lim_{x \to +a} f(x)$.

But Hind did not proceed in this manner. Following Newton (see the appendix 'On the lemmas of Newton cited by Marx') he considered the limit simply the 'last' value of the function of the 'last' value of the independent variable. In other words he looked at $\lim_{x \to +a} f(x)$ as the point of the lower boundary of the values of the function not on the interval $a < x < \alpha$ but on the segment $a \leqslant x \leqslant \alpha$. He assumed the 'last' value $f(a)$ to be already defined; but in that case all of the above procedure loses meaning, since α may take the value a, and to find the lower boundary of all possible values of the function, consisting now of only the one $f(a)$, now becomes that same $f(a)$.

This was just what Marx wanted to say, apparently, when he noted, obviously having in mind Hind's determination, that it is meaningless to treat $3x^2$ as the limit value of the function $3x^2$ as h goes to zero, later terming such treatment a 'well-worn tautology' (see pp.124-6 and notes 90-92); where he calls generally 'childish' and 'the origin of the first mystical and mystifying method of calculus' (see p.126) the actual approach to the limit, the assumption, that the limit value of the function is formed as its 'last' value at the 'last' value of the argument.

This circumstance, that the actual approach to the limit by no means resolves the difficulties surrounding infinitely small quantities, becomes particularly evident in the case when the 'last' value of the independent variable is 'infinity'. So, in particular, if we consider the sequence $\{a_n\}$, then the limit must be that member of the series for which $n = \infty$; so we regard a limit as the end (the last term) of an infinite (that is, without an end) series of terms. It is hardly surprising that this concept of the 'actual limit' should be no clearer than the concept of 'infinitely small quantities' which Marx called 'mystical'.

As is well known, the definition of the limit of a function, not requiring the carrying-out of an infinite number of steps and permitting an exact formulation in terms of only finite variables and parameters, gained currency in mathematics only after the time of Cauchy, that is, in the 70s of the last century. But even at this time the authors of many widely-distributed textbooks did not clearly understand that the limit was not to be interpreted actually; that even in cases where the function is continuous at the point a, that is, the limit of the function $f(x)$ as $x \to a$ is equal to $f(a)$, nevertheless it must be shown equal to $f(a)$ on the condition that, no matter how closely x approaches a, it never reaches it.

With regard to Marx's mathematical manuscripts it is essential for us to note, that if the value $f(a)$ is undefined but the limit $f(x)$ exists as $x \to a$ (corresponding to x over the interval $(a - k, a + k)$) then we may simply predefine the function of $f(x)$ at the point a, $f(a)$, as that limit, by definition. Such a predefinition of the value of the function is also a predefinition of *continuity*. The limit of the function $f(x)$ as $x \to a$ would in this case be the value of the already well-defined function with $x = a$. This however does not mean that one may treat the value $f(a)$ as the determination of the known single-valued function $f(x)$, but on the contrary only as a quantity at the *end* of an infinite progression no matter how closely x approaches a. Indeed, Marx himself obviously had such a predefinition of 'continuity' in mind when he called the limit of the expression $\frac{\Delta y}{\Delta x}$ as $\Delta x \to 0$, the 'absolute minimal expression' of the ratio (see, for example, p. 125); by this he graphically had in mind the limit of this ratio as $\Delta x \to 0$ under the condition that there exists a certain number α, such that for $0 < \Delta x < \alpha$ as Δx decreases so does the ratio $\frac{\Delta y}{\Delta x}$. By means of this definition of a function Lacroix works out the example he gives (see below p.153). But even so far in the construction of mathematical analysis as Lacroix had gone beyond the metaphysical 'principle of continuity' of Leibnitz, which he regarded as a self-evident axiom, nonetheless he did not consider any other definition of function generally possible. Regarding the fact that Marx quite obviously allowed other means of definition of the ratio $\frac{\Delta y}{\Delta x}$ as $\Delta y = \Delta x = 0$, see p.18 and note 18.

We now give some of Hind's own words which may be necessary in reading Marx's manuscripts and from which follow the conclusions set out above.

In his introductory chapter 'On the method of limits' Hind begins with definition number one, to wit:

> 'By the limits of a quantity allowed to vary in value we intend those values, between which are contained *all* those values which it may have throughout *all* its changes; beyond which it may not extend and distinct from which may be made the quantity; — provided that they can be expressed in finite terms' (that is, without the use of the symbols 0 and/or ∞ —*S.A. Yanovskaya*. See Hind, p.1, our italics.)

With this definition there follows a series of examples, in which, however, not once is brought into clear view nor once is demonstrated that the 'limit' spoken of by the author actually fulfils the requirements formulated in Definition One. The first of these examples is the following:

> 'The quantity ax, wherein x admits of all possible values from zero or 0, to infinity, or ∞, becomes 0 in the former case and ∞ in the latter; and consequently the limits of the algebraical expression ax are 0 and ∞: the first is the *inferior*, the second the *superior* limit.' (Sic. It is here obviously assumed that $a > 0$.)

Already the first example plunges the student into confusion. How can the quantity ax be made to differ from the value ∞ by finite quantities, 'a magnitude from which it may be made to differ by quantities less than any that can be expressed in finite terms'? Indeed, following Hind, when x assumes a finite value the difference $\infty - ax$ is equal to infinity, but when $x = \infty$, then $ax = \infty$, and the difference $\infty - \infty$ is undefined.

In the second example (it is necessary to consider, naturally, the values in these conditions of x and a respectively) the lower and upper limits of the expression $ax + b$ are found, appropriately enough, at b and infinity.

In the third example the lower limit of the fraction $\frac{ax + b}{bx + a}$, that is, $\frac{b}{a}$

is found by simple substitution of 0 in the place of x in the expression, and the upper limit, $\frac{a}{b}$, by the substitution of ∞ in place of x in the equivalent fraction $\frac{a+\frac{b}{x}}{b+\frac{a}{x}}$. An explanation of under what conditions the values given to a and b respectively appeared actually in the lower and upper limits does not accompany the example. There is not even a hint of the question of whether if the values are tested they will satisfy the adduced definition of 'limits' (to check, for example, that we are looking at monotonic functions). The reader is thus pre-'prepared' to find a limit to a function through the direct substitution into its expression (or into its re-arranged expression in those cases where the immediately given continuous expression is devoid of any meaning) of the limit value of the independent variable.

The fourth and the sixth examples, exactly those examples which typify point two of the introductory chapter — in which proceeds the gradual 'transition' from the concept of inferior and superior limits of the function to the more conventional concept of limit and in which is revealed the actual character of limit according to Hind — we reproduce here in full. From them it will become sufficiently clear what a jumbled character is attributed to any general account of the concept of limit by this author:

'Example 4: The sum of the geometric series

$$a+\frac{a}{x}+\frac{a}{x^2}+\text{ etc. },$$

$$\frac{a\left(\frac{1}{x^n}-1\right)}{\frac{1}{x}-1}=\frac{ax\left(1-\frac{1}{x^n}\right)}{x-1};$$

now, if $n=0$, the inferior limit is manifestly $=0$; but if $n=\infty$, $\frac{1}{x^n}$ becomes 0, and therefore the superior limit is $\frac{ax}{x-1}$; which is usually called the sum of the series continued *ad infinitum*.

'Example 6. If a regular polygon be inscribed in a circle, and the number of its sides be continually doubled, it is evident that its perimeter approaches more and more nearly to equality with the periphery of the circle, and that at length their difference must become less than any quantity that can be assigned; hence therefore, the circumference of the circle is the limit of the perimeters of the polygons.' (pp.2-3)

Here one no longer speaks of one of the 'limits' of the sequence nor any more about the superior of the limits, as would naturally follow from Definition One, but simply of the limit in the usual sense.

'2. *To prove that the limits of the ratios subsisting between the sine and tangent of a circular arc, and the arc itself, are ratios of equality.*

'Let p and p' represent the perimeters of two regular polygons of n sides, the former inscribed in, the latter circumscribed about, a circle whose radius is 1, and circumference = 6.28318 etc. = 2π; then (trig.)

$$p = 2n \sin \frac{\pi}{n}, \text{ and } p' = 2n \tan \frac{\pi}{n};$$

hence

$$\frac{p}{p'} = \frac{2n \sin \frac{\pi}{n}}{2n \tan \frac{\pi}{n}} = \cos \frac{\pi}{n},$$

and if the value of n be supposed to be indefinitely increased, the value of $\cos \frac{\pi}{n}$ is 1, and therefore $p = p'$; now, the periphery of the circle evidently lies between p and p', and therefore in this case is equal to either of them; hence on this supposition, an nth part of the perimeter of the polygon is equal to an nth part of the periphery of the circle: that is,

$$2 \sin \frac{\pi}{n} = \frac{2\pi}{n} = 2 \tan \frac{\pi}{n}, \quad \text{or} \quad \sin \frac{\pi}{n} = \frac{\pi}{n} = \tan \frac{\pi}{n},$$

or the sine and the tangent of a circular arc in their *ultimate* or *limiting state*, are in a ratio of equality with the arc itself.' (p.3)

The word 'limit' or 'limits' occurs here only in the verbal formulation of the theorem, but recalling that formulation we see that one surmises that the requirement is to show the equality of $\frac{\sin x}{x}$ and $\frac{\tan x}{x}$ as x goes to 0. However, Hind's proof can hardly be considered satisfactory by the standards of his time. Indeed, from the above account it is evident that the author desires to show that

$$\sin\frac{\pi}{n} = \frac{\pi}{n} = \tan\frac{\pi}{n} \text{ as } n = \infty \tag{1}$$

But even here, in order to have $\cos\frac{\pi}{n} = 1$ when $n = \infty$ he already assumes that $\frac{\pi}{n} = 0$ when $n = \infty$, and therefore as well $\sin\frac{\pi}{n} = \sin 0 = 0$ and $\tan\frac{\pi}{n} = \tan 0 = 0$. That is, in order to prove equation (1) — from which, of course, it by no means follows by itself the theorem on the limit of the ratio $\frac{\sin x}{x}$ as $x \to 0$ — the assumptions immediately preceding the introduction by the author of this equation are missing completely.

It remains equally difficult to explain how all this confusing account could possibly demonstrate the superiority of this method of limits, literally interpreted, over the method of infinitely small quantities, in this case simply the identification of an infinitely small segment of the perimeter of the circle with its chord.

In Boucharlat's textbook as well (see p.vii) the method of limits is treated as an improvement on the method of infinitely small quantities: 'repairing that which may be imperfect in this last'. There is, however, no attempt in Boucharlat's course to define what is meant by 'tends to (such-and-such) a limit' (or how to make certain that such-and-such a quantity actually tends toward such-and-such a limit). In it the concept of limit, as well as of 'actual', appears for the first time in evaluating the derivative of the function $y = x^2$. We reproduce here in full that passage which elicited critical remarks from Marx in his manuscript 'On the ambiguity of the terms "limit" and "limiting value".'

'By attending to the second [right-hand] side of equation (2)

$$\frac{y' - y}{h} = 3x^2 + 3xh + h^2 \,, \tag{2}$$

we see that this ratio is diminished the more h is diminished, and that when h becomes 0 this ratio is reduced to $3x^2$. This term $3x^2$ is therefore the limit of the ratio $\frac{y'-y}{h}$, being the term to which it tends as we diminish h.

'Since, on the hypothesis of $h = 0$, the increment of y becomes also 0, $\frac{y'-y}{h}$ is reduced to $\frac{0}{0}$, and consequently the equation (2) becomes

$$\frac{0}{0} = 3x^2 \tag{3}$$

'This equation involves in it nothing absurd, for from algebra we know that $\frac{0}{0}$ may represent every sort of quantity; besides which it will be easily seen, that since dividing the two terms of a fraction by the same number the fraction is not altered in value, it follows that the smallness of the terms of a fraction does not at all affect its value, and that, consequently, it may not remain the same when its terms are diminished to the last degree, that is to say, when they become each of them 0.' (pp.2-3)

For a correct understanding of the above-mentioned manuscript of Marx it is essential to note that in Boucharlat's account the transition from the equation of the form $\frac{\Delta y}{\Delta x} = \Phi(x_1, x)$ (where $y = f(x)$) to an equation of the form $\frac{dy}{dx} = f'(x)$ is presented as divided into those parts to the left and to the right in the first equation above: from $\frac{\Delta y}{\Delta x}$ to $\frac{dy}{dx}$ and from $\Phi(x_1, x)$ to $f'(x)$. And the limit of the ratio $\frac{\Delta y}{\Delta x}$ — corresponding to the $\frac{y_1-y}{h}$ of equation (2) — is evidently considered equivalent to the expression $\frac{0}{0}$, denoted $\frac{dy}{dx}$. So, in his determination of the differential of x, having deduced the equation $\frac{y_1-y}{h} = 1$, Boucharlat concludes: 'Since the quantity h does not enter into the second side of this equation, we see that to pass to the limit it is sufficient to change $\frac{y_1-y}{h}$ into $\frac{dy}{dx}$ which gives $\frac{dy}{dx} = 1$, and therefore $dy = dx$.' (p.6)

The case where the limit appears equal to zero Boucharlat treats as equivalent to the nonexistence of a limit. So, taking the derivative of

$y = b$ and obtaining the equation $\frac{dy}{dx} = 0$, he concludes, 'so there is neither limit nor differential' (p.6).

Boucharlat obtains the limit of the ratio $\frac{\sin x}{x}$ as $x \to 0$ in essentially the same manner as Hind, although in a more intelligible form. He proves at first the theorem given as an example in his textbook, that 'the arc is greater than the sine, and less than the tangent'. (p.24) However, he makes no mention of the fact that immediately follows, viz:

$$\frac{\sin x}{\tan x} < \frac{\sin x}{x} < \frac{\sin x}{\sin x} \left(0 < x < \frac{\pi}{2}\right),$$

that is, that the ratio $\frac{\sin x}{x}$ lies between $\cos x$ and 1. All this aside, following Hind, Boucharlat writes:

> 'It follows from the above, that the limit of the ratio of the sine to the arc is unity; for since, when the arc h . . . becomes nothing, the sine coincides with the tangent; much more does the sine coincide with the arc, which lies between the tangent and the sine; and, consequently, we have, in the case of the limit, $\frac{\sin h}{\text{arc } h}$ or rather $\frac{\sin h}{h} = 1$.' (p.29)

The condition that for $h = 0$ the ratio $\frac{\sin h}{h}$ is 'transformed' into $\frac{0}{0}$, that is, in general, is undefined, and the conclusion drawn on no more ground than 'the sine coincides with the arc' when this last is changed into zero, all these embarrass Boucharlat no more than they embarrass Hind.

We have dwelt long enough, obviously, on the treatment of the concept of limit in the textbooks of Hind and Boucharlat in order to clarify those passages in the manuscript 'On the ambiguity of the terms "limit" and "limiting value"' in which Marx criticised these authors' actual transition to the limit, (concerning which see notes 90-92).

In order to understand other passages of the manuscripts, and in particular Marx's characteristic ratio treatment of the limit, closer to the contemporary one, it is advisable to introduce certain opinions regarding the concept of limit in other sources with which Marx familiarised himself, first of all the 3-volume *Traité* of Lacroix on the differential and integral calculus, 1810.

Following Leibnitz, Lacroix considered all sorts of functions obeying the requirements of the law of continuity, but considered the passage to the limit to be the expression of this law, '*c'est-à-dire de la loi qui s'observe dans la description des lignes par le mouvement, et d'après laquelle les points consécutifs d'une même ligne se succedent sans aucun intervalle.*' (p.xxv) ('that is, the law which is observed of lines when described by [their] movement, and according to which there is not the slightest interval between successive points of the same line'). For any such change in the quantity is impossible to understand without considering its two different values, between which the interval is being considered, since the law of continuity must be expressed in terms of it, that '*plus il est petit, plus on se rapproche de la loi dont il s'agit, à laquelle la limite seule convient parfaitement*', (*ibid*: 'the smaller it becomes the more closely it approaches the law which it obeys, to which only the limit fits with complete agreement'). Lacroix also explains that this role of continuity in mathematical analysis seemed to him appropriate in order to '*employer la méthode des limites*' (p.xxiv) for the construction of a systematic course-book of mathematical analysis.

The concepts 'infinite' and 'infinitely small' Lacroix considers determined only in a negative sense, that is, as '*l'exclusion de tout limite, soit en grandeur, soit en petitesse, ce qui n'offre qu'une suite de négations, et ne sourait jamais constituter une notion positive*' (p.19 'the exclusion of any limit whether of greatness or of smallness, this only offers a series of negations and never rises to constitute a positive notion'). And in a footnote on the same page he adds '*l'infini est necessairement ce dont on affirme que les limites ne peuvent être atteintes par quelque grandeur conçevable que ce soit,*' ('the infinite is necessarily that of which one believes its limits cannot be surpassed by any conceivable quantity no matter how large'). In other words, Lacroix does not accept any actual infinity: neither an actual infinitely large quantity nor an actual infinitely small one.

Lacroix introduces the concept of limit in the following manner:

> 'Let there be given a simple function $\frac{ax}{x+a}$ in which we suppose x to be augmented positively without end. In dividing the numerator and divisor by x the result

$$\frac{a}{1+\frac{a}{x}},$$

clearly shows that the function will always remain less than a, but will approach that value without a halt, since the part $\frac{a}{x}$ in the denominator diminishes more and more and can be reduced to any degree of smallness which one would want. The difference between the given fraction and the value a is expressed

$$a - \frac{ax}{x+a} = \frac{a^2}{x+a},$$

and becomes therefore smaller and smaller as x is larger, and *could be made less than any given quantity, however small*; it follows that *the given fraction can approach a as closely as one would want*: a is therefore the *limit* of the function $\frac{ax}{x+a}$ with respect to the indefinite increase of x.

'The terms which I now am stating comprise the true value [which it is necessary to atrribute to] the word *limit* in order to understand all of what it implies.' (pp.13-14)

Already in Lacroix there is no longer any assumption of a monotonic or piecewise monotonic function, and his limit is not, in general, a one-sided limit: the variable may approach its limiting value in any manner whatsoever. In place of the concept of absolute value Lacroix employs, although not consistently, the expression 'value without sign', the meaning of which, however, remains unspecified. He emphasised that the function may not only attain its limiting value but in general may even pass beyond, to oscillate in its vicinity. But Lacroix still did not formulate in clear terms the restriction on the independent variable that in its approach to its limiting value α, related to the passage to the limit, it is assumed that it does not attain α, that is, that the limit is not to be understood actually. As long as the function with which he is concerned is continuous, that is, its limits coincide with the value of the function at the limiting value of the independent variable, he expresses himself as would a man who believed that the approach of the independent variable to its limiting value must in the passage to the limit be completed by reaching that value.

It must also be noted that Lacroix uses the same one word 'limit' for the designation of the *limit* — an end which as we have seen was conceived by him in a much more general, more precise way, and closer to the contemporary sense than anything in the concepts of the textbooks of Boucharlat and Hind which Marx criticised — as he uses in several instances for the designation of the limit value as well.

These lines on the concept of limit in the long treatise of Lacroix — which, as we know, Marx considered his most reliable source of information on the fundamental concepts of mathematical analysis, such as function, limit etc. — are obviously sufficient to clarify what Marx had in mind when he noted briefly regarding the concept of limit in Lacroix's treatment, that 'this category, brought into general use in [mathematical] analysis largely by Lacroix's example, acquires great significance as a replacement for the category "minimal expression" ' (p.68). It is clear, first of all, that Marx actually understood what he was doing when he introduced, in dealing with the ambiguity of the term 'limit', the concept of the 'absolutely minimal expression', in the same sense as that which we recognise today in the concept of limit. Marx foresaw, it is also clear, that with the concept of limit as understood by Lacroix we are forced, after completely replacing, obviously, the less satisfactory concept of limit, to perform the unnecessary introduction of the special — new — concept of the 'absolutely minimal expression'; in other words, we are faced with the necessity of replacing the latter.

It is probably appropriate, in connection with this same extract from the manuscripts of Marx which we are discussing at the moment, but also with regard to a variety of other passages of the manuscripts, to introduce the words of Lagrange with respect to the concept of limit from the introduction to his *Theory of Analytic Functions* (*Oeuvres Lagrange*, Vol IX, Paris, 1881).

Speaking about the attempts by Euler and d'Alembert to regard infinitely small differences as absolutely zero, with only their ratios entering into calculus, and to see these as the limits of the ratios of finite or indefinitely small differences, Lagrange wrote (p.16):

> '*Mais il faut convenir que cette idée, quoique just en elle-même, n'est pas assez claire pour servir de principe à une science dont la certitude doit être fondée sur l'evidence, et surtout pour être presentée aux commençants.*' ('But it is necessary to admit that this idea, however correct in itself, is not at all clear enough to serve as the principle of

a science whose certitude must be founded solely on evidence and must above all be presentable to beginners.')

Later (p.18) he remarks, in connection with the Newtonian method of the remaining ratios of disappearing quantities, that

> '*cette méthode a, comme celle des limites dont nous avons parlé plus haut, et qui n'en est proprement que la traduction algébraique, le grand inconvénient de considérer les quantités dans l'état où elles cessent, pour ainsi dire, d'être quantité, car, quoiqu'on conçoive toujours bien le rapport de deux quantités, tant qu'elles demeurent finies, ce rapport n'offre plus à l'ésprit une idée claire et precise aussitôt que ses termes deviennent l'un et l'autre nuls à la fois.*' ('This method has, like that of limits of which we spoke earlier and of which it is only the algebraic translation, the great inconvenience of having to consider quantities in the state in which they, so to speak, cease to be quantities; since however well one understands the ratio of two quantities so long as they remain finite, such a ratio no longer presents a clear and precise idea to the understanding unless both of its terms become zero simultaneously.')

Lagrange then turned to the attempts of 'the clever English geometrician' [John] Landen to deal with these difficulties, attempts which he valued highly, although he considered Landen's method too awkward. (See Appendix IV, 'John Landen's *Residual Analysis*', pp.165-173)

Of himself, Lagrange wrote that already in 1772 he maintained 'the theory of the development of functions into a series containing the true principles of differential calculus separate from all consideration of infinitely small quantities or of limits'. (p.19)

Thus it is clear that Lagrange considered the method of limits no more perfect than the method of infinitely small quantities and that this was related to his understanding that the limit of which one speaks in analysis is understood actually as the 'last' value of the function for the 'last' ('disappearing') value of the independent variable.

APPENDIX II

ON THE LEMMAS OF NEWTON CITED BY MARX

On a separate sheet attached to his draft sketch of the course of historical development of mathematical calculus, Marx referred to the Scholium of Lemma XI of Book One and the Lemma II of Book Two of Newton's *Principia*, devoted to two fundamental concepts used by Newton throughout his mathematical analysis, the concept of 'limit' and 'moment'.

In the commentary (*scholium*) to Lemma XI of the first book to *Principia mathematica de philosophiae naturalis* Newton attempts to explain the concept of 'ultimate (limiting) ratio' and 'ultimate sum' by means of a not very transparent comparison: 'a metaphysical, not mathematical assumption,' Marx characterised it. Indeed, Newton writes:

> 'Perhaps it may be objected, that there is no ultimate ratio of evanescent quantities; because the ratio before the quantities have vanished, is not the ultimate, and when they are vanished, is none. But by the same argument it may be alleged that a body arriving at a certain place, and there stopping, has no ultimate velocity; because the velocity, before the body comes to the place, is not its ultimate velocity; when it has arrived, there is none. But the answer is easy; for by the ultimate velocity is meant that with which the body is moved, neither before it arrives at its last place, and the motion ceases, nor after, but at the very instant it arrives; that is, that velocity with which the body arrives at its last place, and with which the motion ceases. And in like manner, by the ultimate ratio of evanescent quantities is to be understood the ratio of the quantities not before they vanish, nor afterwards, but with which they vanish. In like manner the first ratio of nascent quantities is that with which they begin to be. And the first or last sum is that with which they begin and cease to be (or to be augmented or diminished). There is a limit which the velocity at the end of a

motion may attain, but not exceed. This is the ultimate velocity. And there is a like limit in all quantities and proportions that begin and cease to be.' (Sir Isaac Newton's *Mathematical Principles of Natural Philosophy*, transl. Andrew Motte, rev. Florion Cajori, Berkeley, Univ. of Calif. Press, 1934, pp.38-39)

In present-day mathematics 'the velocity of a body at the given moment t_0' is defined with the help of the mathematical concept of limit, and the use by science of such a definition may lead to a variety of considerations, including those of an ontological character. However, the scientific definition of the velocity of a body at a given moment by means of a certain limit of the ratio of vanishing quantities can serve neither as a demonstration of the existence of such a limit nor, *a fortiori* as a justification for the definition of this limit as 'the ratio of the quantities not before they vanish, nor afterwards, but with which they vanish,' that is, as some sort of ratio of zeroes, the value of which is somehow compared to the speed which a body must have at the very moment when it reaches a place where its movement ends. Clearly, however, from such a 'definition' it is impossible to extract by mathematical calculations any corresponding limit, and we are essentially in a logical circle: *velocity at the moment* t_0 is factually described as a certain *limit*, *the limit*, itself, however, is then defined by means of the *velocity at the moment* t_0, the existence of which in this case now really seems to be some sort of 'metaphysical, not mathematical, assumption'.*

Lemma II of the second book of *Principia mathematica* contains the following explanation of the concept of 'moment' (or infinitely small):

'I understand . . . the quantities I consider here as variable and indetermined, and increasing or decreasing, as it were, by a continual motion or flux; and I understand their momentary increments or decrements by the name of moments; so that the increments may be esteemed as added or affirmative moments; and the decrements as subtracted or negative ones. But take care not to look upon finite particles as such. Finite particles are not moments, but the very quantities generated by the moments. We are to conceive them as the just nascent principles of finite magnitudes. Nor do we in this Lemma regard the magnitude of the moments,

* Consisting in that the reflection is understood as the reflected object: the contemplation in our thoughts of the anticipated goals of abstract mathematical concepts is understood as the real existence of the ideal object. —*Ed.*

but their first proportion, as nascent. It will be the same thing if, instead of moments, we use either the velocities of the increments and decrements (which may also be called the motions, mutations and fluxions of quantities), or any finite quantities proportional to those velocities.'

It is natural that this explanation — in which Newton once again employs a 'metaphysical, not mathematical assumption', this time with respect to the existence of differentials ('moments') — should have interested Marx first of all.

But this lemma might also have attracted his attention insofar as in it Newton attempts to show the formula for the differentiation of the product of two functions without resorting to the suppression of the infinitesimals of higher order.

This (unsuccessful) attempt proceeds in the following way: Let $A-\frac{1}{2}a$ be the value of the function $f(t)$ at the point t_0, $B-\frac{1}{2}b$ be the value of the function $g(t)$ at the same point t_0, and a and b increments of the respective functions f and g on the interval $[t_0, t_1]$. (Lower we denote these Δf and Δg respectively.) Then the increment of the product $f(t).g(t)$ on the segment $[t_0, t_1]$ is:

$$\left(A+\frac{1}{2}a\right)\left(B+\frac{1}{2}b\right)-\left(A-\frac{1}{2}a\right)\left(B-\frac{1}{2}b\right),$$

that is, $Ab+Ba$, which Newton also understood as the differential ('moment') of the derivative of the functions f and g at t_0. But here $Ab+Ba$ is not $f(t_0)\Delta g+g(t_0)\Delta f$, but

$$\left(f(t_0)+\frac{1}{2}\Delta f\right)\Delta g+\left(g(t_0)+\frac{1}{2}\Delta g\right)\Delta f,$$

that is, different from $f(t_0)\Delta g+g(t_0)\Delta f$ by the same quantity of $\Delta f.\Delta g$ whose suppression Newton wanted to avoid. Identifying, although implicitly, $Ab+Ba$ with $f(t_0)\Delta g+g(t_0)\Delta f$, however Newton in fact employed such a suppression.

As is apparent from the first drafts of the piece on the differential (see, for instance p.76), Marx at first wanted to elucidate the historical path of the development of differential calculus by the use of the example of the history of the theorem of the derivative. Therefore it is not surprising that Lemma II should have drawn Marx's attention in this connection.

Since the textbooks from which Marx made extracts do not specifically refer to Lemma XI of Book One or Lemma II of Book Two of the *Principia*, there is every reason to believe that Marx selected them, having already immediately rejected Newton's work.

Since the definition of the limit of the ratio of vanishing quantities by means of the velocity of a body at a given moment t_0 contains no means for the calculation of this limit, Newton actually employs for the performance of such calculation, rather than this definition, certain hypothetical properties of limits sufficient to reduce the calculation of the limits of ratios of vanishing quantities to the calculation of the limits themselves, the numerical value of which is supposed to be completely and rigorously defined. Newton states these hypothetical properties first of all in Lemma I of the first section of Book One of *Principia*: 'The method of first and last ratios of quantities, by the help of which we demonstrate the propositions that follow.' In his notes on the history of differential calculus Marx refers to this lemma together with the scholium to Lemma XI (see pp.75 and 76).

Lemma I states: 'Quantities, and the ratios of quantities, which in any finite time converge continually to equality, and before the end of that time approach nearer to each other than by any given difference, become ultimately equal.' (Newton's *Principia* revised by Florion Cajori, Univ of Calif. Press, 1934, p.29)

However, in the demonstration of this limit the existence of a limit as actually reached *at the end* of the period of time in question is implicitly assumed. Actually, the demonstration is composed of a denial that the value of the quantities obtained 'at the end of this time' can be distinguished from each other.

Thus, *limit* is always understood by Newton in an actual sense and therefore hardly surpasses — in mathematical precision and validity — Leibnitz's actually infinitely small *differentials* and their corresponding *moments*, which, as is well known, Newton used in practice.

APPENDIX III

ON THE CALCULUS OF ZEROES OF LEONHARD EULER

In order to understand those places in the manuscripts of Marx at which the ratio $\frac{dy}{dx}$ is regarded as a ratio of zeroes, at times equal to the value of the derivative of y with respect to x for all values of x and at the same time something which can be treated as an ordinary fraction — where, for example, the product $\frac{du}{dv} \cdot \frac{dv}{dx}$ equals the 'fraction' $\frac{du}{dx}$, 'cancelling' the dv's — it is essential to have an acquaintance with Euler's attempt to construct the differential calculus as a calculus of zeroes. This attempt deserves interpretation as well in view of the fact that Marx specifically refers, in the list of literature appended to his first draft of the history of differential calculus, to chapter III of Euler's *Differential Calculus*, and that Marx calls Euler's account of the calculus 'rational'.

The *Differential Calculus* by the great mathematician and member of the St Petersburg Academy of Sciences Leonhard Euler was published by the St Petersburg Academy in 1755. The basis for this work lies in the attempt to regard differentials as at the point of equalling zero in quantity, yet at the same time as different from zero: a zero with a 'history' of its origin, with various designations (dy, dx and so on) and allowed to be evaluated so that the ratio $\frac{dy}{dx}$ where $y = f(x)$, is distinguished by the fact that it is the derivative $f'(x)$ and can be treated as an ordinary fraction.

Euler undertook this attempt in order to free mathematical analysis from the necessity of treating differentials as actually infinitely small quantities with a clearly contradictory character (appearing to be in some sense zero and non-zero simultaneously). The assertion that 'pure reason supposedly recognises the possibility that the thousandth part of a cubic foot of substance is devoid of any extent', Euler considers 'completely inadequate' (in the sense of 'inadmissible', in context, see the translation [in Russian] of L. Euler, *Differential Calculus*. Moscow-Leningrad, 1949, p.90).

> 'An infinitely small quantity is no different from a vanishingly small one, and thus exactly equal to zero. This includes the definition of infinitely small differentials according to which they are smaller than any given quantity. Actually, if the quantity is to be so small that it is smaller than any possible given quantity, then it could not possibly be not equal to zero; or if it is not equal to zero, then there is a quantity to which it is equal, contrary to the supposition. Thus, if one asks, what is the infinitely small quantity in mathematics, we answer, that it is exactly equal to zero. Consequently, this removes the mystery which is usually attributed to this concept and which for many makes the calculus of infinitely small quantities rather suspicious.' (p.91)

Since the simple identification of the differential with zero did not yield the differential calculus, Euler introduces 'various' zeroes, establishing for them two types of equality, the 'arithmetic' and the 'geometric'. In the 'arithmetic' sense all zeroes are equal to each other, and for any non-zero a, $a + 0$ is always equal to a independently of the 'sort' of zero which is added to a. In the 'geometric' sense of the word, two zeroes are equal only if their 'ratio' is equal to unity.

Euler did not clarify what he understands by the 'ratio' of two zeroes. It is only clear that he attributes to this 'ratio' the usual character of a ratio of non-zero quantities and that in practice by the ratio of two 'zeroes' — dy and dx — he intends the same as that which is expressed in modern mathematical analysis by the term $\lim\limits_{\Delta x \to 0} \frac{\Delta y}{\Delta x}$, for Euler's theory of zeroes does not free mathematical analysis from the necessity of the introduction of the concept of limit (and the difficulties attending this concept).

Since for Euler zero becomes various zeroes (and in the 'geometric' sense they are not even equal to one another), it is necessary to use a variety of symbols. 'Two zeroes', writes Euler, 'may have any geometric ratio to each other, while from the arithmetic point of view their ratio is the ratio of equality. Therefore, since zeroes may have any ratio between them, in order to express these different ratios different symbols are used, especially when it is necessary to determine the geometric ratio between the two different zeroes. But in the calculus of infinitely small quantities nothing larger is formed than the ratio of various infinitely small quantities. Unless we employ different signs for their designation everything will be an enormous mess and nothing would be distinguishable.' (p.91)

If, following this interpretation of dx and dy as 'different' zeroes, the ratio of which is equal to $f'(x)$, we replace $\frac{dy}{dx} = f'(x)$ with $dy = f'(x)dx$, then we have an equation the left and right sides of which will be equal both in the 'arithmetic' sense and in the 'geometric' sense. Actually, the left and right will contain various 'zeroes', but all 'zeroes', as already noted, are equal in the 'arithmetic' sense. Only insofar as the ratio of dy to dx is *completely* equal to $f'(x)$ — that is, both in the 'arithmetic' and 'geometric' senses [the ratio $\frac{dy}{dx} : f'(x)$, where $y = f(x)$, is considered unity even if $f'(x) = 0$] and if the 'ratio' of zeroes is understood correctly as the usual operation of ratio, then we have

$$dy : f'(x)dx = \left(\frac{dy}{dx}\right) : f'(x) = 1 ,$$

or, in other words, dy and $f'(x)dx$ are also equal in the 'geometric' sense.

Obviously, Marx had in mind just this 'complete' equivalence of the equation $\left(\frac{dy}{dx}\right) = f'(x)$ with that of $dy = f'(x)dx$ in the sense not only of the possibility of transition from each of them to the other but also of the treatment of this (and with the strength of this) 'ratio' of 'differential parts' dy and dx as a usual ratio (as a fraction), whatever the quality of the 'differential parts' dy and dx as zeroes ('various' zeroes, variously designated), when he transformed the first of these equations into the second (see *ibid*, p.147).

For a more detailed account of the Euler zeroes and a history of the ideas related to it the reader may consult the article, A.P. Yushkevich 'Euler und Lagrange über die Grundlagen der Analysis', in *Sammelband zu Ehren des 250 Geburtstages Leonhard Eulers*, Berlin, 1959, pp.224-244.

Here we are limiting consideration to two considerations of Euler which are helpful in reading the manuscripts of Marx. The first concerns the concept of the differential as the principal part of the increment of the function. This concept, which plays an essential role in mathematical analysis, particularly in its foundations, Euler introduces in the following way: 'Let the increment ω of the variable x become very small, so that in the expression [for the increment Δy of the function y of x, that is; in] $Pw + Qw^2 + Rw^3 +$ etc.* the terms Qw^2, Rw^3 and all higher orders become so small that in an expression not demanding a great degree of precision they may be neglected compared to the first term Pw. Then, knowing the first differential Pdx, we also know, admittedly approximately, the first difference, that will be Pw; this has frequent use in many cases in which analysis is applied to practical tasks' (p.105, *ibid*). In other words, having replaced in the differential function y of x (that is, in Pdx, where P is the derivative of y with respect to x) the differential dx, equal to zero according to Euler, with the finite [non-zero] increment w of the variable dx, we obtain the very concept of the differential as the principal part of the increment of the function, the starting point of modern-day courses of mathematical analysis.

The analogous concept of the differential as the principal part of the increment of the function is also in the manuscripts of Marx (see the account in manuscript 2768, p. 297 [Yanovskaya, 1968]).

The second consideration concerns the question of the choice of designations specific to differential calculus, that is, of differentials and derivatives. Here interest arises first of all from the fact that Euler interprets the dot designations of Newton as symbolic of the differential, but not the derivative. In fact he writes, 'the name "fluxions" first used by Newton for the designation of speed of growth, was by analogy carried over to the infinitely small increments which a quantity assumes when it as it were varies' (p.103). And similarly later, 'The differentials which they [the English] called "fluxions", they marked with dots which were placed above the letters, so that $\dot{y}$ meant for them the first fluxion of y, $\ddot{y}$ the second fluxion, $\dddot{y}$ the third fluxion and so on.'

This manner of designation, however, did not satisfy Euler, and he continues: 'Although this means of designation depends upon an arbitrary rule, the designation need not be rejected if the number of dots is not large, for they are easily indicated. If, however, it is required

* The *Differential Calculus* of Euler begins with the calculus of finite differences and the theorem which states that 'if the variable quantity x assumes an incremental value w, then the consequential value of the increment of any function of x can be expressed as $Pw + Qw^2 + Rw^3 + \ldots$ etc., which expression is either finite or continues infinitely.' (*Ibid*, p. 103, see also p.61) The proof of this theorem is based on the fact that the class of functions considered by Euler consists of power functions: polynomials and elementary transcendental functions expanded into infinite power series which he treats as if they were finite polynomials — *Ed.*

to write many dots, this method gives rise to a great deal of confusion and inconvenience. In fact, the tenth differential, or tenth fluxion, is extremely inconvenient to indicate thus: $\overset{\substack{\cdot\\\cdot\\\cdot\\\cdot\\\cdot\\\cdot\\\cdot\\\cdot\\\cdot\\\cdot}}{y}$ where by our means of designation, $d^{10}y$ is given easily. There arise occasions when it is necessary to express differentials of much higher, and even infinite, degree; on those occasions the English method of designation is not at all appropriate.' (pp. 103-104)

About the analogous identification (in several instances) by Newton and his followers of the 'fluxions' $\dot{x}$, $\dot{y}$ and so on, with the 'moments' (that is, the differentials) $\tau\dot{x}$, $\tau\dot{y}$, and so on (where τ is an 'infinitely small period of time') Marx also spoke, when he noted (p.78) 'τ plays no role in Newton's analysis of the foundations of functions and therefore may be ignored', and that Newton himself voluntarily neglected τ (*loc.cit.*). Marx used the same expressions, speaking of the method of Newton, as 'the differential of y or $\dot{y}$, of u or $\dot{u}$, of z or $\dot{z}$'. (see p.79)

We must note in addition that Marx primarily emphasised the Leibnitzian symbology of the differential calculus over the symbology of Newton and his followers (see p.94).

APPENDIX IV

John Landen's *Residual Analysis*

Notice of Marx's intention to acquaint himself with the works of John Landen in the British Museum is evident at several places in the mathematical manuscripts of Marx (see p.33).

Marx saw in Landen a possible precursor of Lagrange, attempting to 'rebuild on strictly algebraic lines the foundation of differential calculus' (p.113), and he proposed that the Landen method should be compared to the method Marx categorised as 'algebraic differentiation', but he himself doubted that Landen really understood the essential difference between this method and any other. To convince himself of the truth of this proposal Marx wanted to study in the Museum Landen's *Residual Analysis*.

In the sources available to him Marx could find two earlier opinions of this book: in Hind's textbook (p.128, 2nd ed.) and in Lacroix's long '*Treatise*' (Vol.I, pp.239-240) — which are in fact almost identical since Hind had essentially translated into English the appropriate passage from Lacroix. In Hind we read: 'The notion of establishing this kind of calculus [that is, differential calculus] upon principles purely algebraical, seems however to have originated with Mr John Landen, a celebrated English mathematician who flourished about the middle of the 18th century. In what is termed his *Residual Analysis*, the first object is to exhibit the algebraical development of the difference of the same functions of the quantities x and x' divided by the difference of the quantities themselves, or the development of the expression $\frac{f(x')-f(x)}{x'-x}$, and afterwards to find what is called the *special value* of the result when x' is made $= x$ and when therefore all trace of the divisor $x'-x$ has disappeared.' (And in Lacroix, '. . . and when this quotient $[f(x')-f(x))/(x'-x)]$ is obtained in order not to conserve any trace of the divisor $x'-x$, one sets $x' = x$, since the final goal of the calculation is to arrive at a special value of the above ratio.')

Marx apparently did not succeed in his intention to study Landen's book in the British Museum. An analysis of the contents of the book, however, completely confirms Marx's expressed opinion, which he himself considered 'highly probable'.

The complete title of the Landen book is '*The Residual Analysis, a new branch of the algebraic art, of very extensive use, both in pure mathematics and natural philosophy*. Book I. By *John Landen*. London. Printed for the author, and sold by L.Haws, W.Clarke and R.Collins, at the Red Lion in Paternoster Row, 1764.'

The preface begins with the words:

> 'Having some time ago stumbled across a new and easy method of investigating the binomial theorem with the help of a purely algebraic process, I turned to see whether the means used to investigate this theorem might be of service with other theorems, and I soon found that a certain type of calculation founded on this method may be used in many researches. I call this special method *Residual Analysis*, since in all problems where it is used the basic tools which we employ to obtain the desired result are those quantities and algebraic expressions which mathematicians call residuals.'

Later the author criticises the fluxions calculus of Newton and the differentials of Leibnitz as based on the introduction into mathematics of undefined new 'principles'. Those applied in the calculus of fluxions of Newton he considers the explanation of the significant new terms introduced into the theory, such as the not really existent but nonetheless apparent (as self-evident) concepts, *imaginary motion* and *graphically continuous flow*, which do not belong in any mathematics of clear and distinct ideas but do continue to speak for example of such things as the *speed of time, the velocity of velocity* and so on as unnecessary in the proof (and therefore on the other hand serve as the means of definition of several exact mathematical concepts). In the analysis of Leibnitz he considers undefined the introduction, under cover of new 'principles', of *infinitely small quantities* and the quantity *infinitely smaller than any infinitely small quantity*, the suppression of which (when it is not a matter of accepted approximate results) is: 'a very unsatisfactory (if not erroneous) method to rid us of such quantities' (p.IV). Landen believed that mathematics had no need of such alien principles and that his *Residual Analysis* 'does not require any principles other than those accepted since antiquity in algebra and geometry', 'no less (if not more) in use, than the calculus of fluxions or differential calculus' (p.IV).

The starting-point of residual analysis is in the formula

$$\frac{a^r - b^r}{a - b} = a^{r-1} + a^{r-2}b + \dots + b^{r-1} \qquad (1)$$

(where r is a positive whole number) with the help of which and the formulae* derived from it

$$\frac{v^{\frac{m}{r}} - w^{\frac{m}{r}}}{v - w} = v^{\frac{m}{r}-1} \frac{1 + \frac{w}{v} + \left(\frac{w}{v}\right)^{2} + \ldots + \left(\frac{w}{v}\right)^{m-1}}{1 + \left(\frac{w}{v}\right)^{\frac{m}{r}} + \left(\frac{w}{v}\right)^{\frac{2m}{r}} + \ldots + \left(\frac{w}{v}\right)^{\frac{(r-1)m}{r}}} \qquad (2)$$

$$\frac{v^{-\frac{m}{r}} - w^{-\frac{m}{r}}}{v - w} = - v^{-1}.w^{-\frac{m}{r}} \frac{1 + \frac{w}{v} + \left(\frac{w}{v}\right)^{2} + \ldots + \left(\frac{w}{v}\right)^{m-1}}{1 + \left(\frac{w}{v}\right)^{\frac{m}{r}} + \left(\frac{w}{v}\right)^{\frac{2m}{r}} + \ldots + \left(\frac{w}{v}\right)^{\frac{(r-1)m}{r}}} \qquad (3)$$

(where m and r are positive whole numbers), Landen obtains the derivative of the power function x^p for whole and fractional (positive or negative) values of p as a 'special value' of the ratio

$$\frac{x^p - x_1^p}{x - x_1}$$

at $x = x_1$. In other words, he predefines the ratio $\frac{x^p - x_1^p}{x - x_1}$ at $x = x_1$ as that which fulfils the equality of formulae (1), (2) and (3).

---

* In order to show (2) using (1) it is sufficient to note that

$$\frac{v^{\frac{m}{r}} - w^{\frac{m}{r}}}{v - w} = \frac{v^m - w^m}{v - w} : \frac{v^m - w^m}{v^{\frac{m}{r}} - w^{\frac{m}{r}}} = \frac{v^m - w^m}{v - w} : \frac{(v^{\frac{m}{r}})^r - (w^{\frac{m}{r}})^r}{v^{\frac{m}{r}} - w^{\frac{m}{r}}}.$$

Formula (3) follows easily from Formula (2)

$$\frac{v^{\frac{m}{r}} - w^{\frac{m}{r}}}{v - w} = \frac{(vw)^{\frac{m}{r}}(v^{\frac{m}{r}} - w^{\frac{m}{r}})}{(vw)^{\frac{m}{r}}(v - w)^{\frac{m}{r}}} = - \frac{w^{\frac{m}{r}} - v^{\frac{m}{r}}}{(vw)^{\frac{m}{r}}(w - v)}$$

The 'special value' of the ratio $\frac{y-y_1}{x-x_1}$, where $y = f(x), y_1 = f(x_1)$, at $x = x_1$, Landen designates $[x-y]$.

He obtains the transition to the irrational powers in his examples, beginning with the determination of the 'special value' of the ratio $\frac{v^{4/3}-w^{4/3}}{v-w}$ at $v = w$ (the derivative of $v^{4/3}$ with respect to v) by two different means, one employing formula (2) with $m = 4$ and $r = 3$, the other by the same formula, but 'since $\frac{4}{3} = 1.333\ldots$' using the pairs ($m$ = 13,333, r = 10,000), (m = 133,333, r = 100,000), and so on. Landen saves himself from the difficulties attending this infinite process by remarking that the 'final value' of

$$\frac{1+1+1+1+\ldots(13{,}333 \text{ times})}{1+1+1+1+\ldots(10{,}000 \text{ times})}$$

is obviously equal to $\frac{4}{3}$, the quantity from which [the number] 1.333 . . . is derived (p.7).

After this he makes the transition to the case where $\frac{m}{r} = \sqrt{2} =$ 1.4142 . . . , treating it by means of the second method, that is, as he himself notes, 'approximately', but such that it can in any case be made more 'closely approximate', he again concludes that the 'final value' of

$$\frac{1+1+1+1+\ldots(14{,}142\ldots \text{ times})}{1+1+1+1+\ldots(10{,}000\ldots \text{ times})}$$

'is equal to $\sqrt{2}$, the value from which [the number] 1.4142 etc. is derived (by the taking of the root).' (p.8)

It is not surprising that Landen cannot construct his *Residual Analysis* without employing in one form or another the concept of *limit*. However, in practice he speaks of the limit from the viewpoint of Newton, treating the limit as the 'final value' (as the end) of an infinite (that is, without having any end) sequence. Naturally he did not in fact use this definition, but he approached by this means an approximate evaluation of the point and of the convergence (or divergence) of the process of their sequential values, which prompted the concrete contents of the question to him.

Like other mathematicians of his time, Landen considers it possible to employ freely divergent series in formally structured expressions of infinite series if the former only play an intermittent role in the construction. If a series had to express the value of some sort of quantity which was subject to calculation, then in order for it to be used it had to converge. Landen did not consider it necessary to explain precisely what he had in mind for 'convergent' or 'divergent' series but instead, having expanded (by means of some sort of formative arrangement) the function into a series, he usually points out the radius of convergence of the derived series and introduces methods by which to 'improve' the convergence (to replace the series with another which converges 'more rapidly' to the same limit). Landen thus, among the number of 'principles' 'already accepted since Antiquity in algebra and geometry', obviously includes some concepts of the passage to a limit, with which he deals in practice (when speaking of an approximate calculation, for example). But he had no general concept of 'convergence' or 'limit'. Nor did he have methods for calculating limits (or proving their non-existence) which included a wide variety of classes of functions. Landen therefore looked for a definition of the derivative (the 'special value') which would contain within itself its own algorithm.

Just like Newton, he spoke in terms of the function of x as an analogue of the concept of real numbers. In detail, just as any real number can be regarded as the (finite or infinite) sum of powers to the base 10, of which each one is denoted by the figures 0, 1, 2 . . . 9, so any function of x, according to Newton, ought to be represented as the (finite or infinite) sum of powers of the base x, with each denoted by numbers (coefficients) — that is, as a power series. (A series was considered 'representing' a certain function given in terms of a finite 'algebraic' expression if the series is obtained by formal manipulation from the given function. So, for instance, the series $1+x+x^2+\ldots+x^n+\ldots$ was considered to 'represent' the function $\frac{1}{1-x}$ since it can be obtained by the division of 1 by $1-x$ by means of the division of the polynomial.) The task of finding the derivative of the function $f(x)$ could be represented as equivalent to the analogous task for the power x^p and to the task, once knowing the derivatives of the elements (or factors), of finding the derivative of the sum. Just these problems Landen solved first of all in his *Residual Analysis*. The extension of these methods into functions of several variables and into partial derivatives of various orders, accompanied by a host of technical difficulties, Landen dealt with by means of occasionally very clever formal calculations.

In this it is usually implicitly assumed that the power series corresponding to the function is *single-valued*, that is, if two power series are to represent one and the same function of x, then the coefficients for each of the powers on them must be equal (hence the widespread use of the so-called 'method of undefined coefficients').

As an example illustrating Landen's use of these methods we present his proposed (with several more precise definitions in use even today) demonstration of the binomial theorem of Newton for the general case of a binomial raised to a real exponent. Since Marx devoted special attention to this theorem of Newton, primarily with respect to the theorems of Taylor and MacLaurin (see for example pp. 109, 116), Landen's proof may provide interest in this connection.

Let

$$(a + x)^p = A_1 + A_2 x + A_3 x^2 + \ldots, \qquad (1)$$

where p is any real number and A_1, $A_2 \ldots$ are undefined coefficients assumed to be independent of x. Letting $x = 0$ on both sides of the equation yields $A_1 = a^p$. The differentiation of the complete equation (1) with respect to x (Landen, of course, did not speak of the derivative with respect to x but of the corresponding 'special value' which he had for Ax^r where A is independent of x and r is real) becomes

$$p(a + x)^{p-1} = A_2 + 2A_3 x + 3A_4 x^2 + \ldots \qquad (2)$$

Multiplying equation (1) by p and equation (2) by $(a + x)$, we obtain

$$p(a + x)^p = pA_1 + pA_2 x + pA_3 x^2 + \ldots, \qquad (1')$$

$$p(a + x)^p = aA_2 + \left.\frac{2aA_3}{A_2}\right\}x + \frac{3aA_4}{2A_3}\Bigg\}x^2 + \ldots, \qquad (2')$$

from which, recalling the assumed single valuation of the expansion of the expression $p(a + x)^p$ into a series of powers of x, we have

$$aA_2 = pA_1, \text{ implies } A_2 = \frac{p}{a}A_1 = pa^{p-1},$$

$$2aA_3 + A_2 = pA_2, \text{ implies } A_3 = \frac{p-1}{2a}A_2 = \frac{p(p-1)}{2}a^{p-2},$$

$$3aA_4 + 2A_3 = pA_3, \text{ implies } A_4 = \frac{p-2}{3a}A_3$$

$$= \frac{p(p-1)(p-2)}{2.3}a^{p-3},$$

. . . .

and therefore

$$(a+x)^{p} = a^{p} + \frac{p}{1}a^{p-1} + \frac{p(p-1)}{1.2}a^{p-2}x^{2} + \frac{p(p-1)(p-2)}{1.2.3}a^{p-3}x^{3} + \ldots,$$

which is the binomial theorem of Newton.

Although the residual analysis of John Landen did not become an everyday working instrument among mathematicians — Landen's notation was cumbersome and he (perhaps therefore) did not reach the theorems of Taylor and MacLaurin — it does not follow that Landen's work was generally without influence in the development of mathematics. Landen himself writes (p.45) that several of his theorems from the *Residual Analysis* have 'struck the attention of Mr De Moivre, Mr Stirling, and other eminent mathematicians'. In his *Traité* (Vol 1, p.240) Lacroix agrees that he employs the Landen method as an '*imitation a l'algèbre*' for the proof of the binomial theorem and the expansion of exponential and logarithmic functions into a series. Lacroix's textbook enjoyed a widespread popularity among mathematicians.

However, Lacroix's notice was drawn to Landen through the influence of Lagrange, whose *Théorie des fonctions analytique* Lacroix made the basis for his *Traité*. In the introduction of this book, speaking of the difficulties remaining in the fundamental concepts of analysis according to Newton, Lagrange writes: 'In order to avoid these difficulties, a skillful English geometer having made an important discovery in analysis, proposed to replace the method of fluxions, which until then all English geometricians used consistently, with another method, purely analytical and analagous to the method of differentials, but in which, instead of employing differences of variable quantities which are infinitely small or equal to zero, one uses at first the different values of these quantities which are then set equal, after having made, by division, the factor disappear which this equality sets equal to zero. By this means one truly avoids the infinitely small and vanishing quantities; but the results and the application of this calculus are embarrassing and inconvenient, and one must admit that this means of rendering the principles of calculus more rigorous at the same time sacrifices its principal advantages, simplicity of method and ease of operation.' (In addition to the *Residual Analysis* Lagrange also cites 'the discourse on the same subject published . . . in 1758. See *Oeuvres des Lagrange*, Vol. IX, Paris, 1881, p.18).

The last comment of Lagrange is obviously related to the fact that Landen uses an extremely awkward notation and did not obtain the differential and the operations with the differential symbols of calculus.

Separate from Lagrange, Lacroix concludes that the method of Landen 'reduces essentially to the method of limits' (*Traité*, p.XVII).

APPENDIX V

THE PRINCIPLES OF DIFFERENTIAL CALCULUS ACCORDING TO BOUCHARLAT

Of the books of mathematical analysis available to Marx, obviously of the greatest significance for the understanding of his manuscripts is the textbook of Boucharlat, *Elementary Treatise on the Differential and Integral Calculus*, with which Marx was acquainted in the English version of the third French edition, translated by Blakelock and published in 1828.

This textbook enjoyed a great popularity and was several times reprinted. Its eighth edition with the commentaries of M.H. Laurent, saw the light in Paris in 1881. It was translated into a variety of foreign languages, among them Russian.

Graduate of the Ecole Polytechnique, professor of 'transcendental' (higher) mathematics, author of a series of textbooks of mathematics and mechanics, Jean-Louis Boucharlat (1775-1848) was at the same time a poet, and since 1823, professor of literature at the Parisian Atheneum.

No doubt his literary accomplishments and clarity of exposition were responsible in no small part for the popularity of Boucharlat's textbook. It is clear that Marx did not turn his attention accidentally to the course-book of Boucharlat.

All the same, despite the pretentions of the author to great rigour in his account and to having perfected the 'algebraic' method of Lagrange by means of the method of limits (see the introduction to the fifth edition, 1838, p.VIII) the mathematical level of this course was not very elevated. Even in the fifth (of 1838) and not only in the third edition, the English translation of which Marx consulted, the concepts of limit, function, derivative, differential are introduced thus:*

---

* Marx not only made extracts of this textbook in several of his manuscripts and polemicised with the author regarding the foundations of his methodological essay, but also invested a great deal of effort in the factual examination of the former. Therefore we could hardly do without an acquaintance with the contents of this textbook. Here we produce in detail the contents of the first twenty paragraphs of the course of Boucharlat.

'1. One variable is said to be a function of another variable, when the first is equal to a certain analytical expression composed of the second; for example, y is a function of x in the following equations:

$$y = \sqrt{a^2 - x^2}, \quad y = x^3 - 3bx^2, \quad y = \frac{x^2}{a}, \quad y = b + cx^3 .$$

. . . .

'3. Let us take also the equation

$$y = x^3 \tag{1}$$

and suppose that when x becomes $x + h$, y becomes y', we have then

$$y' = (x + h)^3$$

or, by expanding,

$$y' = x^3 + 3x^2h + 3xh^2 + h^3 ;$$

if from this equation we subtract equation (1) there will remain

$$y' - y = 3x^2h + 3xh^2 + h^3 ,$$

and by dividing by h,

$$\frac{y' - y}{h} = 3x^2 + 3xh + h^2 . \tag{2}$$

'Let us look at what this result teaches us: $y' - y$ represents the increment of the function y when x receives the increment h, because this difference $y' - y$ is the difference between the new state of the value of the variable y and its original state.

'On the other hand since the increment of the variable x is h, it follows from this that the expression $\frac{y' - y}{h}$ is the ratio of the increment of the function y to the increment of the variable x. Looking at the second term of equation (2), we see that this ratio

---

Paragraphs which are specific to the course and particularly those towards which Marx directed critical remarks are reproduced in full. Passages in the manuscripts for whose understanding an acquaintance with these paragraphs is necessary are accompanied by citations to the pages of the Appendices on which the contents of the paragraph are reproduced.

decreases together with the decrease of h and that when h becomes zero this ratio is transformed into $3x^2$.

'Consequently the term $3x^2$ is the limit of the ratio $\frac{y'-y}{h}$; it approaches this term when we cause h to be decreased.

'4. Since, in the hypothesis that $h = 0$ the increment of y also becomes zero, then $\frac{y'-y}{h}$ is transformed into $\frac{0}{0}$, and therefore there is obtained from equation (2)

$$\frac{0}{0} = 3x^2 . \qquad (3)$$

'There is nothing absurd in this equation, since algebra teaches us that $\frac{0}{0}$ may represent any value at all. On the other hand it is clear that since division of both parts of a fraction by one and the same number does not change the value of the fraction, we may then conclude that the smallness of the parts of a fraction has no effect at all on its value, and that consequently it may remain the same value, even when its parts attain the last degree of smallness, that is, are transformed to zero.

'The fraction $\frac{0}{0}$ which appears in equation (3) is a symbol which has replaced the ratio of the increment of the function y to the increment of the variable x; since no trace remains in this symbol of the variable, we will represent it by $\frac{dy}{dx}$; then $\frac{dy}{dx}$ will remind us that the function was y and the variable x. But this dy and dx will not cease to be zero, and we will have

$$\frac{dy}{dx} = 3x^2 . \qquad (4)$$

$\frac{dy}{dx}$, or more precisely its value $3x^2$, is the differential coefficient of the function y.

'Let us note that since $\frac{dy}{dx}$ is the sign representing the limit $3x^2$ (as equation (4) shows), dx must always be located beneath dy. However, in order to facilitate algebraic operations it is permitted to clear the denominator in equation (4), and we obtain $dy = 3x^2dx$. This expression $3x^2dx$ is called the differential of the function y.' (pp. 1-4)

In §§ 5-8 Boucharlat finds dy in the examples

$$y = a + 3x^2, \quad y = \frac{1-x^3}{1-x}, \quad y = (x^2 - 2a^2)\,(x^2 - 3a^2) .$$

In all these cases the expression for the increased value of y, that is (in Boucharlat's notation) for y', is equal to $f(x+h)$ — if $y = f(x)$ — and is represented in the form of a polynomial, expanded in powers of h (with coefficients in x), after which the ratio $\frac{y'-y}{h}$ is easily represented as a polynomial of the same type. Setting $h = 0$ in this ratio gives $\frac{dy}{dx}$, and multiplication by dx completes the search for the expression for the differential dy.

'9. The expression dx is itself the differential of x; let $y = x$, then $y' = x + h$, consequently $y' - y = h$, and then $\frac{y'-y}{h} = 1$. Since the quantity h does not even enter the second term of this equation, it is enough to change $\frac{y'-y}{h}$ to $\frac{dy}{dx}$ which will give $\frac{dy}{dx} = 1$; consequently, by our hypothesis, $dy = dx$.

'10. We find in the same way that the differential of ax is adx; but if we had $y = ax + b$ we also would have obtained adx for the differential, whence it follows that the constant b, unaccompanied by the variable x, provides no term at all upon differentiation or, in other words, has no differential at all.

'In addition one may note that if $y = b$, then in the case before us, where a is zero in the equation $y = ax + b$ and where therefore $\frac{dy}{dx} = a$ is now reduced to $\frac{dy}{dx} = 1$, there is neither limit nor differential.' (p.6)

We see from the above that according to Boucharlat:

1) There is neither a definition of limit, nor of derivative or differential. All these concepts are explained only in examples, and only such that the ratio $\frac{f(x+h)-f(x)}{h}$ is represented as a polynomial expanded in powers of h, with coefficients in x. The evaluation of the limit of this ratio as $h \to 0$ is treated as the supposition that $h = 0$ in the obtained polynomial. Here questions whether there exist other cases, whether in such cases it is possible to 'differentiate', and if so, how, do not even arise.

2) The passage from the derivative $\frac{dy}{dx} = \varphi(x)$ to the differential $dy = \varphi(x)dx$ is regarded as an unlawful operation, carried out only in order to 'facilitate' algebraic calculation.

3) From the fact that for $h \neq 0$

$$\frac{f(x+h)-f(x)}{h} = \varphi(x, h)\,, \qquad (A)$$

is drawn the conclusion that for $h = 0$, that is, when $\frac{f(x+h)-f(x)}{h}$ loses all meaning, $\left(\text{is transformed into } \frac{0}{0}\right)$, equation (A) retains significance, that is, we should obtain

$$\frac{0}{0} = \varphi(x, 0)\,. \qquad (B)$$

In other words, it is considered $\varphi(x, h)$ should be defined (and continuous) for $h = 0$ and that equation (B) follows logically from equation (A) — although the expression $\frac{0}{0}$ is without meaning.

4) The limit or differential equalling zero is rationalised as indicating that 'there is neither limit nor differential' although at the same time dy and dx are always zeroes (if $\varphi(x) \neq 0$, then the differential, equal to $\varphi(x) \,.\, 0$, exists, if $\varphi(x) \equiv 0$, then it doesn't).

It is not surprising that such a treatment of the fundamental concepts of the differential calculus did not satisfy Marx. And in fact the first of his outlines of the opening paragraph of the course-book of Boucharlat (see p.65 of the present edition) contains critical remarks concerning that author. But Marx was displeased in particular with the fact that the fundamental concept of differential calculus — the concept of the *differential* — appeared without foundation and its introduction justified only because it 'facilitates algebraic operations'. (see the manuscript 'On the Differential', p.15).

In §11 of Boucharlat's book the remark is made, 'sometimes the increment of the variable is negative; in that case we must put $x - h$ for x, and proceed as before'. In the example $y = -ax^3$ by this means is obtained $dy = -3ax^2dx$, and the conclusion drawn: 'We see that this comes to the same thing as supposing dx negative in the differential of y calculated on the hypothesis of a positive increment.' But for Boucharlat dx is 0. The question of the meaning of 'negative zero' never came into his head, however. (In the works of this period there was still no general concept of 'absolute value'.)

Since the following three paragraphs, §§12-14, are particularly characteristic of Boucharlat's course-book and since they are related to a variety of passages in the manuscripts of Marx, the text of these paragraphs is reproduced here in full.

'12. Before proceeding further, we must make one essential remark; viz., that in an equation, of which the second side is a function of x, and which for that reason, we will represent generally by $y = f(x)$, if on changing x into $x + h$, and arranging the terms according to the powers of h, we find the following development:

$$y' = A + Bh + Ch^2 + Dh^3 + \text{ etc. }, \qquad (C)$$

we ought always to have $y = A$.

'For if we make $h = 0$, the second side is reduced to A. In regard to the first side, since we have accented y only, to indicate that y has undergone a certain change on x becoming $x + h$, it follows necessarily, that when h is 0, we must suppress the accent of y and the equation will be reduced then to

$$y = A.$$

'13. This will give us the means of generalising the process of differentiation. For, if in the equation $y = f(x)$ in which we are supposed to know the expression represented by $f(x)$, we have put $x + h$ in place of x; and after having arranged the terms according to the powers of h, are able to obtain the following development:

$$y' = A + Bh + Ch^2 + Dh^3 + \text{ etc.}$$

or rather, according to the preceding article,

$$y' = y + Bh + Ch^2 + \text{ etc. },$$

we shall have

$$y' - y = Bh + Ch^2 + \text{ etc. },$$

therefore

$$\frac{y' - y}{h} = B + Ch + \text{ etc.}$$

and taking the limit, $\frac{dy}{dx} = B$; which shows us that the differential coefficient is equal to the coefficient of the term which contains the first power of h, in the development of $f(x + h)$, arranged according to ascending powers of h.

'14. If instead of one function y, which changes its value in consequence of the increment given to the variable x which it contains, we have two functions, y and z, of that same variable x, and we know how to find separately the differentials of each of these functions, it will be easy, by the following demonstration, to determine the differential of the product zy of these functions. For if we substitute $x + h$ in place of x, in the functions y and z, we shall obtain two developments, which, being arranged according to powers of h, may be represented thus,

$$y' = y + Ah + Bh^2 + \text{ etc. }, \tag{5}$$

$$z' = z + A'h + B'h^2 + \text{ etc.} \tag{6}$$

Passing to the limit, we shall find

$$\frac{dy}{dx} = A\ , \qquad \frac{dz}{dx} = A'\ ; \tag{7}$$

multiplying equations (5) and (6) the one by the other, we shall obtain

$$\begin{aligned} z'y' = zy + Azh + \ & Bzh^2 + \text{ etc.} + \\ + A'yh + \ & AA'h^2 + \text{ etc.} + \\ & + B'yh^2 + \text{ etc.}\ , \end{aligned}$$

therefore

$$\frac{z'y' - zy}{h} = Az + A'y + (Bz + AA' + B'y)h + \text{ etc }\ ;$$

and taking the limit, and indicating, by a point placed before it, the expression to be differentiated, we shall get

$$\frac{d\,.zy}{dx} = Az + A'y\ ;$$

and suppressing the common factor dx,

$$d\,.zy = zdy + ydz\,.$$

'Thus, to find the differential of the product of two variables, we must multiply each by the differential of the other, and add the products.' (pp.6-8)

In §15 this is correctly used to determine the differential of the product of three variables, in §16 to obtain the differential of the fraction $\frac{y}{z}$.

In § 17 the differential of the power function $y = x^m$ for a positive m is obtained from the formula

$$\frac{d\,.xyztu\text{ etc.}}{xyztu\text{ etc.}} = \frac{dx}{x} + \frac{dy}{y} + \frac{dz}{z} + \frac{dt}{t} + \frac{du}{u} + \text{ etc.} \qquad (9)$$

under the supposition that x, y, z, t, u etc. are equal to x and are taken m times.

§ 18 contains the formula for correctly differentiating a power function.

In § 19 by the use of the formula for operation with the differential symbols (having related the problem to previous cases) it is correctly shown in the cases of fractional and negative exponents.

In § 20 the differential of a power [function] is obtained immediately by the expansion of $(x + h)^m$ according to the binomial theorem of Newton.

In the third edition of Boucharlat's course-book, the English translation of which Marx used, there is a 'Note Second' in the appendices with a title beginning, 'Considerations which prove the solidity of differentiation . . . ' Since this comment attracted Marx's special attention, its text is introduced here (in part):

> 'With the exception of the differentials of circular functions, which, as we have already seen, are readily found by the formulae of trigonometry, all the other monomial differentials, such, for example, as those of x^m, a^x, $\log x$, etc., have been deduced from the binomial theorem alone. We have, it is true, had recourse to the theorem of MacLaurin, in the determination of the constant A in the exponential formulae, but we might have dispensed with it.'

Later, with the help of formal manipulations of infinite series which are not at all well-founded from the modern point of view, it is shown how this might be done, after which Boucharlat concludes:

> 'It follows from this that the principles of differentiation rest all of them on the binomial theorem alone, and since that theorem has been demonstrated, in the elements of algebra, with all the rigour possible, we may conclude that our principles are founded on a firm basis.' (p.362)

Thus it is clear that Boucharlat adhered to the viewpoint of the 'algebraic' differential calculus of Lagrange, which he tried to improve with the help of the concept of limit. His 'improvement', however, reduced to the fact that whereas Lagrange wanted to avoid the application of the then not yet well-based concept of limit and simply defined the derivative of $f(x)$ as the coefficient of the first power of h in the expansion

$$f(x+h) = f(x) + Ah + Bh^2 + Ch^3 + \dots, \qquad (1)$$

where $A, B, C, \dots$ are functions of x, Boucharlat 'uncovered' the same derivative ('differential coefficient') by means of the passage to the limit, which last, however, consisted simply of taking $h = 0$ in the expression

$$\frac{f(x+h) - f(x)}{h} = A + Bh + Ch^2 + \dots, \qquad (2)$$

which is derived purely formally from equation (1). Boucharlat gave no definition of the concept of 'limit' or any sort of commentary on it. He limited himself to hints to the effect that the limit is the last value of the unlimitedly close approach (that is *not having* a last value) of a variable quantity. No wonder that such a concept of limit could not possibly satisfy Marx.

APPENDIX VI

TAYLOR'S AND MACLAURIN'S THEOREMS AND LAGRANGE'S THEORY OF ANALYTIC FUNCTIONS IN THE SOURCE-BOOKS USED BY MARX

1) These theorems, including Lagrange's closely connected theory of analytic functions, attracted Marx's particular attention, and he specifically devoted a series of longer, more important manuscripts to them (see mss 4000, 4001, 4300, 4301, 4302 [not translated]). In order to understand these manuscripts, particularly the critique to which Marx subjected the proof of Taylor's theorem which had been introduced in the handbooks at Marx's disposal, it is necessary to become acquainted with these proofs and with the corresponding ideas of Lagrange. Before we approach them, however, let us establish something of the history of Taylor's and MacLaurin's theorems.*

Taylor's Theorem is actually included as the 7th proposition of the book *Methodus incrementorum directa et inversa* by the English mathematician Brook Taylor (1685-1731), published in London in 1715. Taylor had already advised his teacher John Machin by letter of this result in 1712. 'Taylor's Theorem' was so called for the first time in 1784 in the article 'Approximations' in the French Encyclopaedia (*Encyclopédie méthodique*) of Condorcet. In 1786 Simon Lhuilier also used this title in the book *Exposition élémentaire des calculs supérieure*, honoured by an award by the Berlin Academy of Sciences (the thesis had been offered in a competition of the Academy). Since that time

---

* As sources we have used: M. Cantor, *Vorlesungen über Geschichte der Mathematik*, 2nd ed, Vol.3, pp.378-382; D.D. Mordukhai-Boltovskoi, 'Kommentarii k "Metodu raznosteei" ' (Commentary on the 'Method of Differences') in the book *Isaak Nyuton, Matematicheskie roboty*, Moscow/Leningrad 1937, pp.394-396; M.V. Vygodskii, 'Vstupitel'noe slovo k "Differentsial'nomu ischisleniya" L. Eilera' (Introduction to L. Euler's 'Differential Calculus') in the book L. Euler, *Differentsial'noe ischislenie*, Moscow/Leningrad, 1949, pp.10-12; G. Vileitner, *Istoriya matametiki ot Dakarta do serediny XIX stoletiya*, Moscow 1960, pp.138-140; O. Becker & J.E. Hofmann, *Geschichte der Mathematik*, Bonn, 1951, pp.200-201, 219; G.G. Tseiten, *Istoriya matematiki v XVI i XVII vekakh*, Moscow/Leningrad, 1938, pp.412, 445; D.Ya. Stroik (Dirk Struik), *Kratkii ocherk istorii matematiki* Moscow 1964, no.153-154. For more complete coverage see the book by M. Cantor, pp.378-382.

the theorem has entered all the handbooks of mathematical analysis and no one has called it anything else. We know nowadays, however, that the Scottish mathematician James Gregory already possessed it in the years 1671-72.

Both Gregory and Taylor approached 'Taylor's Theorem' starting from finite differences. At this point Taylor addressed himself directly to the problem of considering Newton's deliberately utterly vague explanation of his interpolation formulae. Newton had obtained his theorem by first allowing the independent variable to differ from zero by a (finite) increment and then — after a series of transformations — returning it to zero 'by dividing it into an infinitely large number of pieces'. If we replace Taylor's extremely cumbersome notation by more modern notation, the proof appears as follows.

Let $y = f(x)$, where x is a variable which is varied, as he says, 'uniformly', that is, obtaining the successive values x, $x+\Delta x$, $x+2\Delta x, \ldots, x+n\Delta x = x+h$. And let the corresponding values of $f(x)$ be y (or y_0), $y_1, y_2, \ldots, y_n$. Let the successive differences (differences of the first order) between y_{k-1} and y_k ($k = 0, 1, \ldots, n-1$) be $\Delta y, \Delta y_1, \ldots, \Delta y_{n-1}$; the differences between these differences (differences of the second order) are $\Delta^2 y, \Delta^2 y_1, \ldots, \Delta^2 y_{n-2}$; and so on. In order to visualise all this, let us write it in schematic form:

$$\begin{array}{ccccccccccc}
x & & x+\Delta x & & x+2\Delta x & & x+3\Delta x \ldots & & & & x+n\Delta x \\
y & & y_1 & & y_2 & & y_3 & \ldots & & & y_n \\
 & \Delta y & & \Delta y_1 & & \Delta y_2 & & \ldots & & \Delta y_{n-1} & \\
 & & \Delta^2 y & & \Delta^2 y_1 & & & \ldots & \Delta^2 y_{n-2} & & \\
 & & & \Delta^3 y & & & & \ldots\ \Delta^3 y_{n-3} & & & \\
\cdot & \cdot & \cdot & \cdot & \cdot & \cdot & \cdot & \cdot & \cdot & \cdot & \cdot
\end{array}$$

It is then clear that:

$$y_1 = y + \Delta y\ ,$$

$$y_2 = y_1 + \Delta y_1\ , \qquad \Delta y_1 = \Delta y + \Delta^2 y\ ,$$

$$y_3 = y_2 + \Delta y_2\ , \qquad \Delta y_2 = \Delta y_1 + \Delta^2 y_1\ , \qquad \Delta^2 y_1 = \Delta^2 y + \Delta^3 y\ ,$$

Hence we further obtain:

$$f(x+\Delta x) \;= y_1 = y+\Delta y\,,$$
$$f(x+2\Delta x) = y_2 = (y+\Delta y)+(\Delta y+\Delta^2 y) = y+2\Delta y+\Delta^2 y\,,$$
$$f(x+3\Delta x) = y_3 = (y+2\Delta y+\Delta^2 y)+(\Delta y+\Delta^2 y)+(\Delta^2 y+\Delta^3 y)$$
$$= y+3\Delta y+3\Delta^2 y+\Delta^3 y\,,$$

. . . .

Having observed the general regularity, Taylor concludes from this that:

$$f(x+n\Delta x) = y+n\Delta y+\frac{n(n-1)}{1.2}\Delta^2 y+\frac{n(n-1)\,(n-2)}{1.2.3}\Delta^3 y + \ldots + \Delta^n y\,, \qquad (1)$$

which is Newton's interpolation formula (for interpolation across equal intervals). Its similarity to Newton's binomial theorem is striking — particularly the fact that the coefficients in the expansion into $\Delta y, \Delta^2 y, \ldots, \Delta^n y$ are exactly the same.

Setting $n\Delta x = h$ (Taylor used v instead of h), we will have:

$$n = \frac{h}{\Delta x},\quad n-1 = \frac{h-\Delta x}{\Delta x},\quad n-2 = \frac{h-2\Delta x}{\Delta x},$$
$$\ldots, n-(n-1) = \frac{h-(n-1)\Delta x}{\Delta x}.$$

Substituting these values for n, $(n-1)$, $(n-2)$, . . . into formula (1), Taylor obtained (in our notation):

$$f(x+h) = y+h\,\frac{\Delta y}{\Delta x}+\frac{h(h-\Delta x)}{1.2}\,\frac{\Delta^2 y}{\Delta x^2} + \frac{h(h-\Delta x)\,(h-2\Delta x)}{1.2.3}\,\frac{\Delta^3 y}{\Delta x^3}+\ldots, \qquad (2)$$

although he didn't even write out the last term ,

$$\frac{h(h-\Delta x)\,(h-2\Delta x)\ldots(h-(n-1)\Delta x)}{1.2\ldots n}\,\frac{\Delta^n y}{\Delta x^n}.$$

He now assumed h to be fixed, n to be actually infinitely large, and Δx to be actually infinitely small ('zero'), inferring that this transformed $\frac{\Delta y}{\Delta x}$ into the first fluxion $\dot{y}$ $\left(\frac{dy}{dx}\text{ according to Leibnitz}\right)$, $\frac{\Delta^2 y}{\Delta x^2}$ into the second fluxion $\ddot{y}$ $\left(\frac{d^2y}{dx^2}\text{ according to Leibnitz}\right)$, and so on. This transforms formula (2) into:

$$f(x + h) = y + \dot{y}h + \ddot{y}\frac{h^2}{1.2} + \dddot{y}\frac{h^3}{1.2.3} + \ldots,$$

that is, into Taylor's series.

Thus, even beginning with finite differences and only then 'removing' them, Taylor still operated strictly in the style of Newton and Leibnitz, with actually infinitely large and actually infinitely small quantities and with the symbolic formulae of the calculus of fluxions, not wondering whether they had any 'real equivalent' and not bothering to consider, of course, the convergence of the obtained series (even to the value of $f(x + h)$). One must note here that, although Taylor was an ardent adherent of Newton's in the quarrel with Leibnitz and therefore never used the latter's notation nor ever cited him, it is nonetheless no accident that Euler presented the proof* in the language of Leibnitz. As D.D. Mordukai-Boltovskoi notes, in essence Taylor addressed the Newtonian fluxions from the Leibnitzian, not the Newtonian, standpoint, namely from that of finite differences (see the *Kommentarii* cited in Yanovskaya, 1968, p.396).

As for the history of MacLaurin's Theorem, it must be noted first of all that it was already present in Taylor in the form of a special case of his theorem at $x = 0$. It is true that, unlike MacLaurin, Taylor never used the 'MacLaurin series' for the expansions already known at this time, for a^x, $\sin \frac{x}{a}$, $\cos \frac{x}{a}$ which are more easily obtained using this theorem.

Furthermore, with respect to the manuscripts of Marx, who specifically mentioned that he borrowed the 'algebraic expansion' directly from MacLaurin, it must be noted that the proofs of MacLaurin's Theorem (by the method of indeterminate coefficients) which were presented in Boucharlat's and Hind's textbooks actually belonged to MacLaurin himself. Such direct borrowing from the author whose

* Still, Euler proved Taylor's theorem following Taylor. See L. Euler, *Differential Calculus*, chapter 3, 'On the Approximation of Finite Differences', §§44-48.

name the theorem bears may also have taken place, of course, with reference to Taylor's Theorem. The bibliographic list which Marx compiled while preparing the historical sketch is apparent evidence that he had decided to become acquainted with Taylor's work in the original, although he did not succeed in carrying out this intention.

2) We find the same order in which Marx criticised the proof of Taylor's Theorem in manuscript 4302, in Boucharlat's textbook as well (J.-L. Boucharlat, *Elémens de calcul differérentiel*, 5th ed., Paris, 1838; Marx apparently had an English translation done from a different edition).

Having stated the problem of successive differentiation in § 30 (pp. 19-20) — where, by the way, after having obtained $6a$ as the third derivative of ax^3 he remarks (p.20), 'here it is no longer possible to differentiate since $6a$ is a constant' — Boucharlat passes to MacLaurin's Theorem (§31, pp.20-21), proving it by assuming the proof of Taylor's Theorem (later proved in §§55-57, pp.34-37).

As was already mentioned, Boucharlat proves MacLaurin's Theorem by following MacLaurin himself. He apparently did not read the latter's work, however. In fact, with respect to the title 'MacLaurin's Theorem', Boucharlat writes, 'this theorem, as G.Peacock has noted, was discovered by G. Stirling in 1717, consequently earlier than MacLaurin used it,' although, as we have already mentioned, MacLaurin fully acknowledged that Taylor already had the theorem.

Boucharlat's proof — which raises not a single question about the correctness of the assumptions made, not to mention the convergence of the series under consideration — we present below in almost literal translation.

'Let y be a function of x; let us expand it in terms of x and assume:

$$y = A + Bx + Cx^2 + Dx^3 + Ex^4 + \text{ etc.}; \tag{16}$$

we obtain, differentiating and dividing by dx:

$$\frac{dy}{dx} = B + 2Cx + 3Dx^2 + 4Ex^3 + \text{ etc.},$$

$$\frac{d^2y}{dx^2} = 2C + 2.3Dx + 3.4Ex^2 + \text{ etc.},$$

$$\frac{d^3y}{dx^3} = 2.3D + 2.3.4Ex + \text{ etc.},$$

. . . .

Let us denote by (y) that into which y is transformed when $x = 0$,

by $\left(\frac{dy}{dx}\right)$ that into which $\frac{dy}{dx}$ is transformed when $x = 0$,

by $\left(\frac{d^2y}{dx^2}\right)$ that into which $\frac{d^2y}{dx^2}$ is transformed when $x = 0$,

. . . .

the preceeding equations give us

$$(y) = A,\ \left(\frac{dy}{dx}\right) = B,\ \left(\frac{d^2y}{dx^2}\right) = 2C,\ \left(\frac{d^3y}{dx^3}\right) = 2.3D\ ,$$

whence we extract

$$A = (y),\ B = \left(\frac{dy}{dx}\right),\ C = \frac{1}{2}\left(\frac{d^2y}{dx^2}\right),\ D = \frac{1}{2.3}\left(\frac{d^3y}{dx^3}\right);$$

substituting these values into (16), we will have

$$y = (y) + \left(\frac{dy}{dx}\right)x + \frac{1}{2}\left(\frac{d^2y}{dx^2}\right)x^2 + \frac{1}{2.3}\left(\frac{d^3y}{dx^3}\right)x^3 + \ \ldots\ ; \quad (17)$$

and this is MacLaurin's formula.'

In the following §§ 32-34 (pp.21-22) expansions are found by means of MacLaurin's formula for

$$y = \frac{1}{a + x}\ , \qquad y = \sqrt{a^2 + bx},\ \ y = (a + x)^{\mathrm{m}}\ .$$

By this means the binomial theorem is derived from MacLaurin's Theorem in the third example. In the first appendix to our 5th edition of Boucharlat's texbook entitled 'Proof of Newton's formulae by means of differential calculus', a direct derivation (by the same method of indeterminate coefficients) is given of Newton's binomial theorem (for positive integer powers) by means of successive differentiation. It appears as follows.

Boucharlat begins with an expansion of $(1 + z)^{\mathrm{m}}$, from which the required expansion for $(a + x)^{\mathrm{m}}$ is obtained by the substitution $z = \frac{x}{a}$. Assume, he says,

$$(1 + z)^{\mathrm{m}} = A + Bz + Cz^2 + Dz^3 + Ez^4 + \ \ldots \quad (1)$$

Setting $z = 0$ he obtains $A = 1$ and consequently

$$(1+z)^m = 1 + Bz + Cz^2 + Dz^3 + Ez^4 + \ldots.$$

Differentiating both sides of this equation with respect to z, he next finds

$$m(1+z)^{m-1} = B + 2Cz + 3Dz^2 + 4Ez^3 + \text{ etc.}$$

Referring to the fact that this equation is valid for any z, Boucharlat sets $z = 0$ and obtains by this means $m = B$. Differentiating once more and again setting $z = 0$, he obtains

$$m(m-1) = 2C,$$

whence he finds

$$C = \frac{m(m-1)}{2},$$

after which he concludes: 'In the same manner all remaining coefficients are determined, and upon substituting their values into equation (1) this equation is transformed to

$$(1+z)^m = 1 + mz + \frac{m(m-1)}{1.2}z^2 + \frac{m(m-1)(m-2)}{1.2.3}z^3 + \text{ etc.'}$$

(pp.491-492).

3) Boucharlat also demonstrates Taylor's Theorem by the method of indeterminate coefficients. In this case he not only assumes that an arbitrary function of many variables may be expanded into a series of powers of any of the variables, but he also considers this expansion unique; that is, that the coefficients of any two such expansions (in powers of one and the same variable) must be equal. This makes it possible to apply the method of indeterminate coefficients.

In order to arrive at this possibility, that is, of comparing the coefficients of two expansions of one and the same function, Boucharlat begins with a lemma which asserts that the derivatives of $f(x+h)$ with respect to x and to h are equal. Since Marx expresses dissatisfaction in manuscript 4302 (see Yanovskaya, 1968, p.540 [not translated]) with the demonstration of this lemma in Boucharlat's course-book, while it is impossible even to understand pp.41-42 (see Note 117 Yanovskaya, 1968 [not translated]) of manuscript 3888 without being acquainted with this proof, we present it here in full.

Devoted to this is §55 (pp.34-35), in which we read:

'If in some function y of x the variable x changes to $x + h$, we then obtain one and the same differential coefficient both when x is the variable while h is constant, and when h is the variable while x is constant.

'If in order to show this we substitute $x + h = x_1$* in place of x in the equation $y = f(x)$, we then have $y_1 = f(x)$; the differential of $f(x_1)$ will then be equal to some other function of x_1, represented by $\varphi(x_1)$, multiplied by dx; consequently, $dy_1 = \varphi(x_1)dx_1$ or if we replace x_1 by its value $x + h$,

$$dy_1 = \varphi(x + h)\, d(x + h) \;.$$

But the only change which the hypothesis that x is variable while h is constant introduces into this differential refers solely to the factor $d(x + h)$, which reduces to dx when x is variable while h is constant; consequently, in this case we have

$$dy_1 = \varphi(x + h)dx \;,$$

whence we obtain

$$\frac{dy_1}{dx} = \varphi(x + h) \;. \qquad (35)$$

'If on the other hand we make x constant while h is variable, the factor $d(x + h)$ then reduces to dh and we will have

$$dy_1 = \varphi(x + h)dh \;,$$

that is,

$$\frac{dy_1}{dh} = \varphi(x + h) \;; \qquad (36)$$

comparing these two values for $\varphi(x + h)$, we obtain

$$\frac{dy_1}{dx} = \frac{dy_1}{dh} \;.\text{'}$$

In the following §56 Boucharlat extends this lemma to derivatives of higher order and in §57 uses it to prove Taylor's Theorem. He

* Although Boucharlat does employ Lagrange's notation for derived function, he designates the increased x and y (i.e. $(x + h)$ and $f(x + h)$) as x' and y'. We have replaced this designation with x_1, y_1.

begins this 'proof' with the following words on what he considers — and as Marx calls it — his 'starting equation' (37), applicable to any function: 'Let y_1 be a function of $x + h$; let us assume that when we develop this function into powers of h we obtain

$$y_1 = y + Ah + Bh^2 + Ch^3 + \text{ etc.}, \tag{37}$$

where $A, B, C, \ldots$ are unknown functions of x which are yet to be determined.'

Differentiating equation (37) with respect to h and with respect to x, and having obtained by this means

$$\frac{dy_1}{dh} = A + 2Bh + 3Ch^2 + \text{ etc.},$$

$$\frac{dy_1}{dx} = \frac{dy}{dx} + \frac{dA}{dx}h + \frac{dB}{dx}h^2 \text{ etc.},$$

Boucharlat then sets the coefficients of corresponding powers of h in the two equations equal to each other, referring to the lemma, and by this means obtains the expressions he needs for the coefficients $A, B, C, \ldots$ of y and its successive derivatives. Marx gives an account of this proof on one occasion in manuscript 3888 (sheets 54-55; pp. 50-51 in Marx's enumeration), where he compares it to the proof of MacLaurin's Theorem presented above. He criticises this proof in manuscript 4302, primarily for a lack of foundation for its initial hypothesis.

The following §§58-61 in Boucharlat's book contain examples of expansions of $f(x + h)$ by Taylor's formula in the case of $f(x)$ equals $\sqrt{x}$, $\sin x$, $\cos x$, $\log x$. Questions about the convergence of the series obtained are not even mentioned. Cases of inapplicability of the Taylor series are only considered in the very last paragraphs of the first part of the book (devoted to differential calculus) which are printed in small type.

The concluding §62 of the section on Taylor's Theorem and its applications is devoted to a proof of MacLaurin's Theorem from Taylor's Theorem. Marx reproduces this proof in full in manuscript 3888 (see sheets 55-56; pp.51-52 in Marx's enumeration).

Notes

NOTES

[The following is a complete, unabridged translation of notes to the 1968 Russian edition (referred to as Yanovskaya, 1968), covering pages 1-139 in this edition. Commentary by the translators is indicated by square brackets — Ed.]

[1] The manuscript was written in 1881 for Engels. This is the first work in a series of manuscripts conceived by Marx and devoted to a systematic exposition of his ideas on the nature and history of differential calculus. In this work he introduces his concepts of algebraic differentiation and the corresponding algorithm for finding the derivative for certain classes of functions. On the envelope enclosing the manuscript there is the notation in Marx's handwriting: 'For the General'. This was Engels's nickname in Marx's family because of his articles on military questions. Having acquainted himself with the manuscript, Engels answered Marx in a letter on 18 August 1881 (see p.xxvii). The published German text of the manuscript reproduces exhaustively Marx's text. Some of the preparatory material (drafts and supplements) is published on page 473 of Yanovskaya, 1968. Variant readings from the unpublished drafts are provided in footnotes. The manuscript was published for the first time (not in full) in 1933 in Russian translation in the collection *Marxism and Science* (*Marksizm i estestvoznanie*), Moscow, Partizdat, 1933, pp.5-11; and in the journal *Under the Banner of Marxism* (*Pod znamenem marksizma*) No.1, 1933, p.15ff. This is the first time it has been published in German.

[2] In order to avoid confusion with the designation of derivatives, Marx's notation x', y', . . . for the new values of the variable has been replaced here and in all similar cases by x_1, y_1, . . .

In the sources which Marx used there was as yet no concept of absolute value. Therefore Marx frequently (apparently in order to be positive) regards only the growth in the value of the variable, but sometimes (see p.109 of this volume and p.514 Yanovskaya, 1968) he speaks also of the 'increase of *x*' in a positive or negative increment *h*'.

[3] In keeping with the accepted terminology of the source-books which Marx consulted, a finite difference is here understood always to be a non-zero difference.

[4] Marx distinguishes in each equation two sides (where now we speak of two parts), the left hand and the right hand which do not always play symmetric roles. On the left-hand side of the equation he frequently places two different, equivalent expressions joined by the conjunction 'or'.

[5] In the mathematical literature which was at Marx's command the term 'limit' (of a function) had no well-defined meaning and was understood most often as the value the function actually reached at the end of an infinite process in which the argument approached its limiting value (see Appendix I, pp.144-145). Marx devoted an entire rough draft to the criticism of these shortcomings in the manuscript, 'On the Ambiguity of the Terms "Limit" and "Limit Value" ' (pp.123-126). In the manuscript before us Marx employs the term 'limit' in a special sense: the expression, given by predefinition, for those values of the independent variable at which it becomes undefined. For Marx, the ratios $\frac{\Delta y}{\Delta x}$ $\left(\text{at } \Delta x = 0 \text{ this is transformed to } \frac{0}{0}\right)$ and $\frac{dy}{dx}$, interpreted as the symbolic expression of the ratio 'of annulled or vanished differences', that is, of $\frac{0}{0}$, are such expressions. With respect to the ratio $\frac{\Delta y}{\Delta x}$, Marx (influenced to a certain degree by the definitions of this concept in Hind and Lacroix; see Appendix I, p.143) took this to be an expression which is identically equal to this ratio when $\Delta x \neq 0$, but which has been predefined by continuity when the ratio is transformed to $\frac{0}{0}$. The 'limit', at that point, consequently, must be the 'preliminary derivative' (concerning which see p.6 and note 7). Exemplifying this, Marx writes (on p.6), with respect to the ratio $\frac{\Delta y}{\Delta x}$ where $y = ax^3 + bx^2 + cx + d$: 'The "preliminary derivative" $a(x_1^2 + x_1 x + x^2) + b(x_1 + x) + c$ appears here as the limit of a ratio of finite differences; that is, no matter how small we allow the differences to become, the value of $\frac{\Delta y}{\Delta x}$ will always be given by this "derivative".' Later (on p.7), Marx mentions that setting x_1 equal to

x, that is, setting $\Delta x = 0$, 'reduces this limit value to its absolute minimum quantity ,' giving its 'final derivative'.

Analogously, by 'the limit of the ratio of differentials' Marx in this manuscript means the 'real' ('algebraic' — see note 6) expression which provides the value for this ratio; in other words, the derived function. Marx writes, however, that in the equation $\frac{dy}{dx} = f'(x)$, 'neither of the two sides is the limiting value of the other. They approach one another, not in a limit relationship, but rather in a relationship of equivalence,' (see p.126). But here, the concept of 'limit' (and of 'limit value') is used in another sense, close to the one accepted today. Marx uses the term 'absolute minimal expression' (see, for example, p.125) in a sense even closer to the contemporary concept of limit, when he writes in another passage (see p.68) that it is interchangeable with the category of limit, in the sense given it by Lacroix and in which it has had great significance for mathematical analysis (for Lacroix's definition, see Appendix I pp.151-153).

6 By 'algebraic' Marx understands any expression which contains symbols neither of the derivative nor of differentials. Such a use of the term 'algebraic expression' was characteristic of mathematical literature at the beginning of the 19th century.

Marx frequently distinguishes between the concepts 'function of (*von*) x' and 'function in (*in*) x', that is, the function as a correspondence and the function as an analytical expression (see p.506 Yanovskaya, 1968). In the manuscript before us he does not adhere to this distinction strictly, speaking most of the time of simply 'the function x (*die Funktion* x [rendered 'the function of x' in English])', perhaps because he always has in mind only functions given by a certain 'algebraic expression'. He provides a correspondence relating the value of the dependent variable y to the value of the independent variable x by means of the equation $y = f(x)$, where y is the dependent variable and $f(x)$ is an analytic expression with respect to the appearance of the variable x in it.

7 The essence of Marx's method of algebraic differentiation consists of his predefinition (for $x_1 = x$) of the ratio of finite differences (having meaning only when $x_1 = x$),

$$\frac{f(x_1) - f(x)}{x_1 - x} \tag{1}$$

by means of continuity. With this goal in mind he writes down the function $\varphi(x_1,x)$, which coincides with (1) for all $x_1 = x$ and which is continuous as $x_1 \to x$. Marx calls such a function $\varphi(x_1, x)$ *the preliminary derived function of the function* $f(x)$, while the function $\varphi(x,x)$, which is obtained from $\varphi(x_1,x)$ under the assumption that $x_1 = x$, he calls the *derivative of the function* $f(x)$. If this function exists (which is a relevant question for the classes of function under consideration), then it coincides with the present-day concept of the derivative, namely:

$$\lim_{x_1 \to x} \frac{f(x_1) - f(x)}{x_1 - x} = f'(x) .$$

Already in Marx's time well-known functions existed for which the operation of differentiation was undefined (see p.117 of the present edition [and note 85, p211]).

[8] Marx reproduces here the formal expansion of the function into a series which is typical of the mathematics books at his command, having left to one side the questions of the series so obtained and the agreement of the value of the function with the limits of the partial sums.

[9] $\therefore$: a symbol employed in the manuscripts to stand for the word 'consequently'.

[10] The text entitled 'Supplementary' comprises the contents of a separate sheet, appended to the manuscript, of independently numbered pages 1 and (on reverse) 2.

[11] By equation of finite differences Marx clearly intends an expression of the form

$$f(x_1) - f(x) = (x_1 - x)\varphi(x_1, x) \text{ . See note 7}$$

[12] At this point S[amuel] Moore wrote in pencil '*Nicht der Fall, diese Factoren sind* $x_1 - x - 1, x_1 - x - 2$ etc.' ('Not the case. These factors are $x_1 - x - 1, x_1 - x - 2$, etc.'.) Obviously Marx intends here not the factors $(x_1 - x)$ but rather the expression $x_1 - x$, and meant to say that the transition to zero of the difference $x_1 - x$, having been preserved in the expression for the preliminary derivative, does not deprive the latter of meaning.

[13] The manuscript dates from 1881. On the envelope attached to the

manuscript is written 'II For Fred' (*II Für Fred*). Marx calls this manuscript the 'second instalment' (see p.33), since, in it he continues to set forth the views at which he arrived in the process of studying mathematics. Engels showed the manuscript to S[amuel] Moore and conveyed the latter's comments to Marx in his letter of November 21, 1882 (see p.xxix). The manuscript 'On the Differential' was first published (not in full) in Russian translation in the 1933 collection *Marxism and Science* (*Marksizm i estestvoznanie*), pp.16-25; and in the journal *Under the Banner of Marxism* (*Pod znamenem marksizma*), 1933, No.1.

[14] Marx thus assumes here that the functions u and z, which, as subsequently becomes clear, are defined by means of the equations $u = f(x)$, $z = \varphi(x)$ (where $f(x)$ and $\varphi(x)$ are expressions 'in the variable x'), are differentiable functions of x. The fact that no further information on the functions $f(x)$ and $\varphi(x)$ is required to prove the theorem on the differential of the product of two functions, is reflected in Marx's graphic comments regarding $\frac{du}{dx}$, $\frac{dz}{dx}$: 'shadow figures lacking the body which cast them, symbolic differential coefficients without the real differential coefficients, that is without the corresponding equivalent "derivative" '(see p.20). Marx also discusses this specifically in his rough draft essays on the differential. Here and hereafter we shall write $d(uz)$ instead of the contraction duz which Marx used in his manuscripts.

[15] The symbols for derivative and differential which are specific to differential calculus are intended here.

[16] In the literature of the 18th-19th centuries the derivative was often called the 'differential coefficient', which is obviously related to the definition of the derivative as the coefficient of the first power of the increment h of the independent variable x in the expansion of the expression $f(x + h)$ into a series of powers of h. The adjective 'real' refers to the fact that the expression for $f'(x)$ contains no symbols which are restricted to differential calculus.

[17] This way of speaking, in which as a result of multiplication by zero 'the variables u and z themselves become equal to zero,' is explained by the fact that in Marx's time there still existed widespread conceptions of mathematical operations on numbers as changing the numbers themselves: the addition of the positive number b to a

'increases the number a', the multiplication of a by 0 'changes the number a to zero', and so on. These conceptions were put on a scientific basis only in the 20th century.

18 The words 'since we can begin the nullification arbitrarily with numerator or denominator' obviously mean that the predefinition of an expression of the form $\frac{f(x)}{g(x)}$, which at $x = a$ becomes $\frac{0}{0}$ and therefore loses any meaning, may be established for $x = a$ in a number of different ways. If we wish to preserve in the predefinition that property of the ordinary fraction which makes it equal to zero when the numerator is equal to zero, then the value of $\frac{f(a)}{g(a)}$ must be zero. 'To begin the nullification with the numerator' in this case simply means to set $\frac{f(a)}{g(a)}$ equal to zero. Since, however, a fraction with a denominator of 0 does not exist, 'to begin the nullification with the denominator' makes it impossible to retain in the predefinition anything of the properties of an ordinary fraction with a zero denominator. But if for all $x \neq a$ $\frac{f(x)}{g(x)} = \varphi(x)$, and $\varphi(x)$ is continuous at the point a (that is, $\lim_{x \to a} \varphi(x) = \varphi(a)$), then it is natural to set $\frac{f(a)}{g(a)}$ equal to $\varphi(a)$, retaining in this manner the equation $\frac{f(x)}{g(x)} = \varphi(x)$ even for $x = a$. If the numerator is also transformed to zero because the denominator is set at zero, then the words 'begin the nullification with the denominator' may be explained naturally as denoting: predefine in the above-mentioned manner, that is, 'using continuity'. In the books which Marx used, even including the large *Traité* of Lacroix, the preservation of the equation $\frac{f(a)}{g(a)} = \varphi(a)$ in the case of $f(a) = g(a) = 0$ was considered independent, in general, of whatever may have been 'derived'; it was a necessary consequence of the metaphysical law of the continuity of 'all real numbers'.

19 There is a slip of the pen here in the text: instead of $x = a$ there appears $x^2 = a^2$. Instead of correcting it, someone, apparently Moore, made insertion marks in the text in pencil, after which he observed, '*und da* $x^2 = a^2 \therefore x = \pm a = = 2Pa$ *oder* 0,' that is 'and since $x^2 = a^2$, then $x = \pm a$, [whence $P(x + a)$] $= 2Pa$ or [=] 0'. Such an interpretation, however, clearly does not agree with the overall context.

[20] Marx here calls the expression $\frac{dy}{dx}$, which was obtained by the transition from a ratio of finite differences to the derivative, the *symbolic differential expression* for $\frac{y_1 - y}{x_1 - x}$, corresponding to $\frac{f(x_1) - f(x)}{x_1 - x}$.

[21] Apparently this concerns the case where the choice of independent variable is not necessarily fixed, where either u or z may be used as the independent variable. In general, if u and z may be considered to be interchangeable functions of one and the same independent variable, then assigning a value to either one of u and z determines the value of the independent variable and, of course, the value of the other function as well. In other words, what is intended here is the invariance of the symbolic operational equation with respect to the choice of independent variable.

[22] Apparently the word '*dir*' (to you) in the phrase '*der dir bekannte*' (which is known to you) was omitted during recopying, although it is preserved in the notebooks. It is to be understood that this concerns the French mathematician L.B. Francoeur, about whom Engels wrote to Marx in the letter of May 30, 1864. The word in quotation marks, 'elegant', refers to Engels's comment, '*Einzelnes ist sehr elegant*' ('Someone is very elegant'), and contains, obviously, a hint of an ironic relationship of Engels to the author under discussion. Francoeur, like Boucharlat and some others, tried to combine the 'algebraic' method of Lagrange (see pp.24) with the differential calculus of Leibnitz, all the while operating with the symbols of differentials. Marx's note of irony about the 'clarity' with which this was done, concerns both Boucharlat and Francoeur. The first, in order to 'facilitate algebraic operation', introduced an absurd formula; the second, suggested that the differential 'appears synonymous to the derivative and differs from it only ambiguously', consequently, he also wrote, 'the derivative of x is $x' = 1$ or $dx = 1$'.

[23] The extract in quotation marks is a text translated from the French of the books of J.-L. Boucharlat. See, for example *Elémens de calcul differential et de calcul intégral* fifth edition, 1838, p.4.

[24] The reduction to its 'absolute minimum' here obviously implies the stated predefinition of the ratio by continuity at $x_1 = x$; that is, in essence, the transition to the limit where $x_1 \to x$.

25 See Appendix III, 'On the Calculus of Zeroes of Leonhard Euler', p.160

26 Marx here makes a distinction between the differential particles (*die Differentiellen*) dx and dy, which represent the 'removed' differences Δx and Δy, and the differential (*das Differential*) dy, which is defined by the equation

$$dy = f'(x)dx \,. \tag{1}$$

This last equation can be treated as an operational formula which makes it possible to find the derivative $f'(x)$ by means of the already determined differentials dy and dx, transforming equation (1) to its equivalent (see note [24])

$$\frac{dy}{dx} = f'(x) \,. \tag{2}$$

27 Marx's argument against applying the method of treatment which already took place in the 'algebraic' differentiation of the simplest functions of first order consists of the following: 1) the step which consists of assuming $x_1 = x$ is superfluous, since the preliminary derivative here already agrees with the final one; that is, that which is specific to the 'algebraic' method of differentiation does not come to light; 2) the extension to the general case of attributes of differential functions of the first order may lead to the completely erroneous conclusion that all derivatives of higher order, beginning with the second, must be equal to zero.

28 That is, consider $\frac{dy}{dx}$ a ratio of infinitely small quantities, as Leibnitz and Newton had done already.

29 That is, to find the derivative of y with respect to x, considering y as a function of x, given by the two equations:

$$1)\ y = 3u^2, \qquad 2)\ u = x^3 + ax^2 \,.$$

30 Marx assumes here that it has already been established that it is correct to operate with differentials as if they were ordinary fractions (see p.24 and Appendix V, p.173).

31 At this point in the manuscript Moore made the following note in pencil: 'On p.12(5) this is proved for the concrete case there investigated. Should it not be proved instead of assumed for the general case also?' [English is garbled in text; recovered from Russian trans-

lation—*Trans.*] This note, however, is based on a misunderstanding. The 'development demonstrated from given functions' consisted of the symbolic expressions $\frac{dy}{du}$ and $\frac{du}{dx}$ which had been obtained as a result of differentiation. Since, as Marx has already assumed, it is correct to operate with such expressions as if they were conventional fractions, the conclusion was natural that

$$\frac{dy}{du} \cdot \frac{du}{dx} = \frac{dy}{dx} \, .$$

32 Marx did not write section III apparently because he did not succeed in carrying out his intention of studying John Landen's book in the British Museum (see Appendix IV).

33 Under this heading are combined three drafts of various sections of the work, 'On the Differential', and several drafts supplementary to it. For more details see pp. 459, 464, 477, 479 of Yanovskaya, 1968.

34 This excerpt is taken from notebooks which Marx entitled 'A. I' and 'B (continuation of A). II' (see pp.459, 464 of Yanovskaya 1968). It begins on the last (unnumbered by Marx) page of the notebook 'A.I' and 'B (continuation of A). II' (see pp.459, 464 of Yanovskaya, 1968). It begins on the last (unnumbered by Marx) page of the notebook 'A.I' and is inserted at various places in the notebook 'B' (Marx distinguished it with special markings). Part of the indicated draft was first published in Russian in 1933 (see *Under the Banner of Marxism* [*Pod znamenem marksizma*] No.1 as well as *Marxism and Science* [*Marksizm i estestvoznanie*], pp.34-43).

35 Marx everyhere calls 'symbolic' (as distinct from 'algebraic'; see note 6) those expressions which contain the symbols specific to differential calculus, *dx*, *dy* etc. He calls 'real' those expressions of functions which do not contain such symbols.

36 The 'operational formulae of differential calculus' here means those symbolic expressions which indicate (see the text below) which operations must be performed on a defined function to obtain the real value of one or another derivative.

37 The notebook 'A.I' ends at this point. At the end of the page is written in Marx's hand, '*Sieh weiter Heft II*, p.9' ('See further notebook II, p.9'). This indicates the notebook 'B (continuation of A)'.

[38] Concerning the characteristics of this type of predefinition by continuity and the possibilities of other predefinitions satisfying these or other conditions, see note 18 and Appendix I, p.146.

[39] That is, when we make the transition from the region of the usual algebra to a function (the dependent variable) for which it is necessary to predefine the ratio

$$\frac{f(x_1) - f(x)}{x_1 - x},$$

which transforms to $\frac{0}{0}$ at $x_1 = x$.

[40] Marx usually calls expressions not containing symbols specific to differential calculus 'algebraic' (see note 6) or 'real' (see note 2). Here and in several other passages he calls them 'actual' (*wirkliche*). Since in Russian mathematical literature the term 'actual' (number) carries another meaning [namely 'real number' — *Trans*], the word 'actual' (expression) is translated as its synonym 'real' [that is in Russian translation; in English 'actual' is not confusing — *Trans*].

[41] The manuscripts of the second and third drafts are in very rough form: they contain many deletions and insertions. The first four pages of the second draft are not preserved, so we begin with the first complete paragraph. These two drafts, less some abridgements, were first published in Russian in 1933 (*Under the Banner of Marxism* [*Pod znamenem marksizma*], No.1, and *Marxism and Science* [*Marksizm i estestvoznanie*], pp.26-34). See 'Preliminary Drafts and Variants of the Manuscript, "On the Differential",' point a, p.477 [Yanovskaya, 1968].

[42] This entire paragraph (beginning with the words 'when the variable quantities increase . . . ') is Marx's German translation of a passage in Hind's book (see T. Hind, 2nd edition, Cambridge, 1831, p.108). The second draft breaks off at this point. The vacant space (more than half a page) which Marx left after this paragraph is apparent evidence that, not finding the desired quotation in Hind, Marx put aside the contemplated research, obviously intending to return to it later.

Material on the differential of a product obtained by the methods of Leibnitz and Lagrange is contained in the text books of Hind and Boucharlat (see Appendix V pp.173) As for Newton's method, the books mentioned do not illustrate it.

[43] The citation is from the book of Boucharlat (see, for example, J.-L. Boucharlat, 7th edition, Paris, 1858, pp.3-4).

[44] Here Marx projects a somewhat different enumeration of the sections of his work from that which he had followed earlier. In Section III he plans to locate materials which in the second draft were located in Section II; in Section IV, to comment on the historical development of differential calculus by means of the example of the history of the theorem on the differential of a product.

[45] In connection with this paragraph see note 5 as well as Appendix I, 'On the Concept of "Limit" in the Sources Cited by Marx', p.151 (where there is a discussion of how in Boucharlat's textbook both sides of the equation $\frac{dy}{dx} = f'(x)$ are treated as limits) and pp.152-153 (where the discussion is about the concept of limit in Lacroix's long *Traité* and Marx's related concept of the word in this paragraph). Exactly what Marx had in mind in his treatment of the symbolic expression as the limit of $f'(x)$ remains unclear. (Perhaps he simply had in mind the fact that the derivative was obtained as a result of the supposition that $x_1 = x$, that is, when the numerator and denominator of the ratio $\frac{\Delta y}{\Delta x}$ both have attained their limit value of zero, so that the expression $f'(x)$ must correspond not to $\frac{\Delta y}{\Delta x}$ but to $\frac{dy}{dx}$.) Regarding Marx's comment on Lagrange's opinion of the concept of limit as understood by Newton, see p.154 as well.

[46] Marx intended to write several supplements to 'On the Differential', four sketches of which survive (for more details see pp.479-490 [Yanovskaya, 1968], which presents a series of extracts from these sketches). Since the drafts are not finished, only two more complete (and understandable) extracts from them are reproduced here. They are adapted from supplements to the second and third drafts.

[47] This is Marx's heading to section A) of the second draft of the supplement to the manuscript 'On the Differential'. Only point 1), containing a short résumé of the basic work on the differential, is published here. The important supplementary material to the latter work here is the direct indication of the geometric applicabiliy of operational formulae. For more detail see p.479 [Yanovskaya, 1968].

48 This is paragraph A) of the third draft of the supplement. The heading is due to Marx. Published here is only point 3), in which Marx (in his characteristically literary style) introduces the application of the theorem of the differential of a product as an operational formula for finding the derivative of a fraction.

49 With his manuscript 'On the Differential', Marx fulfilled a promise to write a specialised piece shedding light on the historical path of the development of differential calculus. In sketches preceding this letter ['On the Differential' was a letter to Engels — *Trans*], he expressed an intention to illustrate the history of differential calculus by means of the history of the theorem on the differential of a product. Obviously Marx succeeded in carrying out neither of these intentions completely. Only the tentative drafts contained in the notebook 'B (continuation of A)', where they alternate with Marx's computations for his work on the differential, have survived. These drafts begin, appropriately for Marx's primary purpose, with an explanation of the methods of Newton and Leibnitz in the example of the theorem on the differential of a product. For the same reason, only the beginning goes like this and not the concluding section explicating the method of d'Alembert. Later Marx passes to a more detailed discussion and critique of the methods of Newton and Leibnitz in general. This brings him to the general periodisation of the history of differential calculus, in which three periods are distinguished: 1) the mystical differential calculus of Newton and Leibnitz, 2) the rational differential calculus of d'Alembert, and 3) the purely algebraic differential calculus of Lagrange, the characterisation of which comprises the second part of the extant drafts of the history of differential calculus. It was this part which Marx apparently decided to develop into a third letter to Engels. The concluding part of the historical drafts presents a more detailed exposition of the general ideas contained in the first part. The drafts are published in full with the exception of notes whose content refers to the work 'On the Differential', which are omitted.

50 The bibliography which Marx presents in this list is accompanied in many cases by indications of the exact passages in the sources cited where the fundamental concepts and methods of differential calculus are discussed. These were not indicated in the textbooks at Marx's disposal. There is therefore every reason to suppose that Marx chose these passages by consulting the corresponding works (in the library

of the British Museum, apparently). The fact that Marx especially distinguished (placed in a panel) the name John Landen is obviously related to the fact that he had decided to acquaint himself particularly well with J.Landen's *Residual Analysis*. For more details on this see Appendix IV. The sources for Marx's notation of the dates of birth and death on the list are unknown. It is only clear that the sources did not have the date of death of Lagrange.

[51] In the scholium (lesson) to Lemma XI of the first book of *Principia Mathematica* and in Lemma II of the second book, Newton explains the fundamental concepts of differential calculus which correspond to our concepts 'derivative' and 'differential'. For more details on these lemmas of Newton see Appendix II, pp.156-159.

[52] See Marx's outlines of these works (with his critical commentaries) on pp.272-280 [Yanovskaya, 1968].

[53] D'Alembert's *Traité des fluides* does not contain any material on the fundamentals of differential calculus. D'Alembert's views on the fundamental concepts of differential calculus were presented in his articles in the *Encyclopédie* and in his *Opuscules mathématiques*. It is not known what attracted Marx's attention to the *Traité des fluides* of d'Alembert.

[54] The third chapter of part one of L[eonhard] Euler's *Institutiones calculi differentialis* deals with the question 'Of Infinity and the Infinitely Small'. For more details see Appendix III. pp160-164.

[55] This book was assembled by the Abbé Moigno 'following the methods and works of Cauchy, published and unpublished'. The first volume of Moigno's *Lectures* appeared in 1840, the second in 1844.

[56] This conclusion (due to Newton) requires clarification: 'since the numerical quantities of all possible magnitudes may be represented as straight lines', the variation of any quantity may be represented as a sort of linear motion of variable velocity. And since during an infinitely small interval of time the speed of motion can be considered to be fixed, then the path, nearly a point, corresponding to this small time interval (of course corresponding also to the variation of our quantity) is equal to the product of this speed (fluxion) and the infinitely small time interval, τ. Therefore 'moments, or infinitely small portions of the quantities generated = the products of their velocities and the infinitely small time intervals'. Regarding the

metaphysical nature of Newton's attempt to provide a basis for the concepts of 'fluent', 'fluxion', and 'moment', corresponding to our 'function', 'derivative', and 'differential', defining them in terms of mechanics, see Appendix II, pp156-157.

[57] It was explained in Note 49 that Marx intended to return to the illumination of the history of the development of differential calculus by means of the example of the history of the theorem on the differential of a product. So he left a vacant space following his unfinished extract from Hind's text. There, after being repeated one more time this section is introduced as an example of the very theorem on the differential of a product in Newton's treatment. (This theorem is introduced as example 3 in Hind's textbook; see Hind, p.109.)

[58] In Hind's textbook Leibnitz's method is not illustrated in the example of the theorem on the differential of a product, so Marx turned to Boucharlat's textbook. This paragraph is an extract from the latter work (see Boucharlat, p.165).

[59] This sentence appears in the extract from Hind's textbook cited above (Hind, p.106). Further on, however, Marx does not introduce the theorem on the differential of a product as developed by Hind. After this text follow five pages in Marx's notebook which have been omitted (pp.16-20). They deal primarily with calculations concerning theorems on the differentiation of fractional and compound functions as well as the solution of problems related to the parabolic curve $y^2 = ax$. We retain only the comments, written at intervals on pp.16-18, in which Marx emphasises the fact that Newton and Leibnitz began immediately with the operational formulae of differential calculus.

Then under the rubric 'Ad Newton' Marx subjects these methods of Newton and Leibnitz to the criticism that all such methods, notwithstanding all the advantages they bring, inevitably imply the introduction of actually infinitely small quantities and their attendant difficulties. Here again the theorem on the differential of a product is used as the basic example.

[60] By $\dot{x}, \dot{y}, \dot{z}$ Newton and his followers usually signified the rate of change (fluxion) of the variables x, y, z (fluents) the derivatives, that is, of x, y, z, with respect to that variable which plays the role of 'time'; by $\tau\dot{x}$, $\tau\dot{y}$, $\tau\dot{z}$ they designated the 'moments' corresponding to the Leibnitzian differentials or infinitely small increments. However, the Newtonians often also used $\dot{x}$, $\dot{y}$, $\dot{z}$ for the 'moments' or differentials. See Appendix III p.160.

61 This discusses the heuristic generalisation where, in the formula

$$\dot{y} = a\dot{x} \,, \qquad (1)$$

y is simply treated as a certain function $f(x)$, while the constant a becomes a new function $f'(x)$ derived from this $f(x)$; according to this formula (1) becomes a special case of the more general formula

$$\dot{y} = f'(x)\dot{x} \,. \qquad (2)$$

Since $\dot{x}$, $\dot{y}$ are treated as increments, even though infinitely small, the factor $f'(x)$ is therefore a function not only of x but also of $\dot{x}$; the 'derived' function $f'(x)$ in formula (2) turns out not to be independent of $\dot{x}$. It is exactly this fact (which compelled the Newtonians to suppress forcibly the terms containing $\dot{x}$, even though the latter must be different from zero for formula (2) to have any meaning) which serves as the basis for the critique of the Newtonian definition of the derivative of the function $y = f(x)$ as the ratio $\frac{\dot{y}}{\dot{x}}$, to which Marx returns several lines below.

62 That is, obtained in the form of a 'real' expression, not containing differential symbols.

63 Several more lines of unclear meaning are omitted.

64 If $\dot{y} = \dot{x}$ and y itself is x, then in order to obtain an equality in which one side does contain the differential symbol $\dot{x}$ it is sufficient simply to divide both sides of the equality $\dot{y} = \dot{x}$ by $\dot{x}$.

65 '*Zuwachs in x*' ('increase in *x*') obviously signifies here a new function in x obtained from the initial function x^2 — in addition to it, so to speak — by means of the binomial theorem: as the coefficient dx in the expansion of $(x + dx)^2$.

66 Obviously this refers to the fact that the immediate result of the application of the binomial is $dy = 2xdx + dx^2$, not $dy = 2xdx$. But the former equality appears to be mathematically correct only as a result of an incorrect premise.

67 The meaning of the expression 'succeeds in two ways' remains obscure. After the colon there follows a point a) without a point b). Perhaps the 'two ways' here are composed of first, the fact that on the left-hand side the fraction $\frac{\Delta y}{\Delta x}$ is transformed into $\frac{dy}{dx}$ (and not identified from the very beginning with $\frac{dy}{dx}$), and second,

the fact that on the right-hand side the terms $3xh + h^2$ are now obtained by means of correct mathematical operations and not by using some sleight of hand.

[68] The expression in quotation marks has been copied from Hind's textbook cited above (§99, pp.128-129).

[69] He obviously has in mind that Taylor's theorem was published in his collection *Methodus incrementorum* in 1715, that is, during the life of Newton, in whose works this theorem does not appear. See in addition Appendix VI p.182.

[70] For material related to the theorems of MacLaurin and Taylor, see pp.109-119 [this edition], 412, 441, 493, 498 [Yanovskaya, 1968].

[71] For Marx's exposition and critique of the fundamental ideas of Lagrange's theory of analytic functions, see p.113 of this edition.

[72] This refers to rough-draft notes, divided into sections, part of which are published in this edition under the general heading 'First Draft'. See pp.76-90 of this edition.

[73] In the manuscripts devoted to the history of differential calculus there are two passages, located almost immediately adjacent to one another, at which Marx proposed to insert: 1) an investigation of the theorems of Taylor and MacLaurin and 2) a discussion of Lagrange's theory of analytic functions (see p.97). Marx did not succeed in accomplishing his intentions, although he had in his possession a great deal of material on these subjects which he had collected from his sources and which served as the foundation from which he arrived at the point of view on the essence of differential calculus which he presented in the works conveyed to Engels. This material is comprised primarily of outlines but also includes manuscripts containing Marx's summarising or critical comments. The most important of these comments are contained in the manuscripts: 1) 'Taylor's Theorem, MacLaurin's Theorem, and Lagrange's Theory of Derived Functions' (for more details see p.441 [Yanovskaya, 1968]) and 2) 'Taylor's Theorem' (unfinished), extracts from which are reproduced here, in order to amplify somewhat Marx's intentions mentioned above. For extracts from other outlines on the same subjects see pp.281, 412 [Yanovskaya, 1968].

[74] In the handbooks on differential calculus at Marx's disposal the derivatives of all elementary functions, except for the trigonometric ones, were actually calculated by means of the binomial theorem.

Marx noted this himself in his manuscript, 'Theorems of Taylor and MacLaurin, First Systematisation of Material' (see pp.419-420 [Yanovskaya, 1968]). Subsequently Marx formulated for this class of function a different means of differentiation which he called the 'algebraic' (see the manuscript 'On the Concept of the Derived Function'). Therefore it is clear that the present manuscript chronologically precedes 'On the Concept of the Derived Function' and 'On the Differential'.

75 Thus, in Hind's textbook (Hind, pp.84-85), after the example containing the derivation of the binomial theorem by means of the expansion of $(x + h)^m$ into the Taylor series there is introduced the derivation of the theorems of Taylor and MacLaurin from the binomial theorem.

76 Here (see also p.514 [Yanovskaya, 1968]) Marx says straight out, that by 'increment' of the value of the variable x he has in mind any change of this value, whether it be a positive or negative increment h.

77 Because, according to Marx, a function in x is a given expression, it represents a combination of symbols which is considered with respect to the appearance in it of the variable x.

In the given case we have before us the terms of the MacLaurin series, that is the product ('combination') of the two expressions, 1) x^k (k = 0, 1, 2, 3 . . .) and 2) its corresponding 'constant function' $\frac{f^{[k]}(0)}{k!}$.

78 Marx calls expressions not containing the variable x 'constant functions' of x. (y), $\left(\frac{dy}{dx}\right)$, $\left(\frac{d^2y}{dx^2}\right)$ and so on are expressions for $f(x)$ and its successive derivatives in which all appearances of the variable x have been replaced by a constant — zero. The result of this substitution in y and in its corresponding derivative $\frac{d^ky}{dx^k}$ is designated in the manuscripts as (y) and, correspondingly, $\left(\frac{d^ky}{dx^k}\right)$. This designation, which Marx borrowed from Boucharlat (see Boucharlat, p.40), has been preserved in this edition.

79 Marx does not explain what exactly he means here by 'the irrational nature of the constant (or variable) function'. Apparently it deals with the fact that in both cases the cause of the origin of 'exceptions' is the presence in the expansion of terms having no rational mathematical meaning: in the first case without any continuity (such as, for exam-

ple, a 'fraction' of the form $\frac{0}{0}$), and in the second without defined values of the variable x (such as, for example, $\frac{c}{x-a}$ at $x = a$). The 'irrationality' of such an expression does not imply that it necessarily contains a radical sign (compare 'algebraic irrationality'), but is used as the opposite of intelligibility (rationality; compare the 'rational differential calculus of Euler and d'Alembert' with the opposite 'mysticism of Newton and Leibnitz'!). Marx gives a short general characterisation of cases of inapplicability at the very end of the manuscript, 'Theorems of Taylor and MacLaurin, First Systematisation of Material' (see pp.440 [Yanovskaya, 1968]).

80 By 'representation in a finite equation' here is obviously meant a representation of the form

$$f(x+h) = P_0 + P_1 h + P_2 h^2 + \ldots + P_n h^n,$$

where n is a positive integer, and $P_i (i = 0, 1, 2, \ldots n)$ are functions of x.

81 For a more detailed exposition of the proof of Taylor's theorem contained in the sources used by Marx, an exposition necessary in order to understand the critique to which Marx subjected it in the following lines, see Appendix VI, p.182.

82 This is an excerpt from the manuscript 'Taylor's Theorem', which is inserted here because it contains in a more concentrated form Marx's viewpoint on the insufficiency of the proof known to him of Taylor's theorem, on its 'algebraic' origin in the binomial theorem, and on its essential difference from the latter (for more details on the unfinished 'Taylor's Theorem' see p.498 [Yanovskaya, 1968]). Since the first paragraph of this extract presents difficulties in reading it in isolation from the preceding text, we note here that in this paragraph Marx summarises the results of the previous section devoted to the critique of the proof of Taylor's theorem in Hind's book. In it (see Hind §74, pp.83-84; §§77-80, pp.92-96):

1) Taylor's theorem is proved under the assumption that the expression $f(x+h)$ may be expanded into a series of the form:

$$f(x+h) = Ph^{\alpha} + Qh^{\beta} + Rh^{\gamma} + \ldots,$$

where $P, Q, R, \ldots$ are functions of the variable x and the exponents $\alpha, \beta, \gamma \ldots$ are increasing positive integers.

2) The 'cases of inapplicability' of Taylor's theorem are considered, with the result that for certain specific values of the variable x these

conditions are not fulfilled (some of the coefficients P, Q, R . . . are not defined — 'do not have finite values' at these points).

3) The attempt is made, following Lagrange, to show that, generally speaking, excluding, that is, certain specific values of the variable x, the conditions under which Taylor's theorem has been proved (the exponents α, β, γ. . . cannot take on negative or fractional values, the functions P, Q, R. . . are not transformed 'into infinity') are fulfilled for any function $f(x)$. After this come Marx's remarks devoted to the insufficiency of this sort of attempt.

83 The words '$x = a$, for example' refer to the example, examined by Hind, of the expansion into a Taylor series of the expression $f(x + h)$ where $f(x) = x^2 + \sqrt{x - a}$. At $x = a$ the expression has the intelligible value $(a + h)^2 + \sqrt{h}$, but the terms of the Taylor series representing it give, according to Hind, only '$a^2 + 2ah + h^2 + 0 + \infty - \infty + \infty -$ etc., not at all defined' (see Hind, p.93).

84 In the function $y = f(x)$, where $y_1 = f(x + h)$ is only the symbolic expression of a binomial of a certain power, one here naturally has in mind the function $y = x^m$, where m is a positive integer.

85 A literal translation of this passage would be, 'which in the course of differentiation can give no result' (*die auf dem Weg der Differentiation kein Resultat liefern können*).

86 Literally: 'in the possible historical part of this manuscript' (*beim etwaigen historischen Teil dieses Manuscripts*).

87 In the manuscript 'On the History of Differential Calculus' Marx notes that from the simple difference in the form of representation of the change in the value of the function originate essential differences in the treatment of differential calculus (see p.102). Regarding this he made reference to the 'introductory pages' in which he developed this thought 'in the analysis of d'Alembert's method' (see *ibid.*) These sheets are of two groups: sheets of one group are marked with the capital Latin letters A to H (see p.471 [Yanovskaya, 1968]), and sheets of the other group with small Latin letters from a to n (see p.498 [Yanovskaya, 1968]).

Since d'Alembert defines the derivative by means of the concept of limit, Marx naturally begins his analysis of the method with a critique of the concept of limit, the inadequacy of which is made clear with the material presented in Appendix I (see 'Concerning the Concept of "Limit" in the Sources Consulted by Marx', p.153). This part of the manuscript occupies sheets A to D (published under the title corresponding to its contents, 'On the Ambiguity of the Terms "Limit" and "Limit Value" '). Also directly related to the above-mentioned passage in the manuscript on the history of differential calculus are sheets E to H, published here under the title, 'Comparison of d'Alembert's Method to the Algebraic Method'. And devoted to essentially the same question are sheets a to g of the other group, which are published here under the title 'Analysis of d'Alembert's Method by Means of yet Another Example'. (For the contents of the remaining sheets of this group see pp.468-470 [Yanovskaya, 1968].) In conformity with Marx's reference to the appended separate sheets devoted to the analysis of d'Alembert's method, they are grouped together here under the general title, 'Appendices to the Manuscript "On the History of Differential Calculus": Analysis of d'Alembert's Method' (pp121-132).

[88] In other words, it is proposed to consider here the expression $3x^2 + 3xh + h^2$ for non-negative values of x and h under the assumption that h tends unboundedly towards zero, remaining different from zero. We recall that in the sources which Marx used there was as yet no concept of absolute value, so that he was not required to consider the sum of all non-negative terms.

[89] Here Marx comes to the basis for his later conclusion, that 'the concept of the limit value may be interpreted wrongly, and is constantly interpreted wrongly' (see p.126), as a consequence of which it is appropriate to replace it by some new term which is unambiguously understandable. As such he proposes the term 'absolute minimal expression', by which is meant the limit in the usual present-day meaning of the word (see p.126 and Appendix I, p.143). Marx's criticism of the 'limit value' defined here and of the way this concept is used in Hind's and Boucharlat's textbooks refers first of all to the fact that the 'limit' is considered there as actual; that is, it is regarded as 'the last' value of the function for 'the last' value of the argument, and therefore represents 'a childishness which has its origin in the first mystical and mystifying method of calculus' (see p.126). In this particular paragraph he obviously has in mind the 'limit value' in the meaning of the definition introduced by Hind (see Appendix I, p.145), who in practice treats it as coinciding with the one-sided limit of a function where the argument approaches a certain number from the right or from the left: in the given case, with the one-sided limit from the right of the function $3x^2 + 3xh + h^2$, considered as a function of h as $h \rightarrow +0$. In contrast to Hind, however, Marx emphasises that this 'limit value' only has meaning if it is not understood as taking place but is calculated with the condition that $h \neq 0$ (here $h > 0$);

that is, he treats it exactly as we do today. At the same time the application of this to the function in consideration, $3x^2 + 3xh + h^2$, does not violate the requirement contained in the definition of 'limit value' (as the exact upper or lower bound to the value of the variable) with which Hind's textbook begins. In fact, as Marx notes, this function firstly, as h approaches zero, constantly approaches its own limit (the lower one, clearly), and secondly, consequently all the more never passes beyond it; that is, it explicitly satisfies both conditions of Hind's definition (Hind himself usually did not verify the satisfaction of these requirements; see Appendix I, p. 145).

90 If the (one-sided) limit of the function $3x^2 + 3xh + h^2$ at the approach of h to zero (from the right; that is, as h decreases) is interpreted actually, that is, the argument h is supposed to attain its limit ('last') value 0, then from the multiplicity of values of the function with respect to which, according to Hind's definition, the limit must be the exact lower bound, it is sufficient to choose the set consisting of only the one value of the function at $h = 0$ (see Appendix I, p. 145), in the given case consequently of only one number $3x^2$ — which, however, as Marx says below, it would be a 'well-worn tautology' to regard as the limit value for $3x^2$ as h approaches zero. In other words, to speak naturally of $3x^2$ as the limit value of $3x^2 + 3xh + h^2$ as h approaches zero at the same time as regarding $3x^2$ as the limit value of $3x^2$ itself as h approaches zero is not intelligible here — most of all because it is in general superfluous: it gives us nothing new.

91 This expression $\frac{0}{0}$ is considered here to be the limit of the quotient $\frac{y_1 - y}{x_1 - x}$, as was done similarly in Boucharlat's textbook (see Appendix I, p. 149), but with the difference that here the limit value (here again in Hind's sense) of the functions $x_1 - x$ and $y_1 - y$ as $x_1 \rightarrow +x$ is not understood by Marx in an actual sense, that is, it remains an assumption that $x_1 \neq x$ (here $x_1 > x$).

92 Here again reference is made to the fact that $\frac{0}{0}$ $\left(\text{or } \frac{dy}{dx}\right)$ is impossible to interpret actually, that is, as the value of the ratio $\frac{y_1 - y}{h}$ at $h = 0$, since in that case, following Hind and obtaining the limiting expression $\frac{0}{0}$ by simply supposing $h = 0$, one would have to admit that the consideration of this expression, in which no trace remains of the ratio $\frac{y_1 - y}{h}$ which contained the variable h, as the limiting value for the

same $\frac{0}{0}$ (regarded as the 'constant' function of h) as $h \rightarrow +0$, in general gives no new result. However, for the expression $\frac{y_1 - y}{h}$ when considered for h distinct from zero (here $h > 0$), it is precisely $\frac{0}{0}$ which, standing opposite the derived function 'as its real equivalent', is, as Marx says, 'its absolute minimal expression', that is, the limit in the usual present-day sense.

93 The original had initially: 'applied in the above differential equations' (*auf obige Differentialgleichungen*), but Marx crossed out the word '*obige*'. It is however clear that here as previously, this does not concern equations in the proper sense of the word, but rather the fundamental formulae of differential calculus having the form of equalities.

94 Here Marx wrote, '. . . to the geometric', a clear slip of the pen.

95 As already noted, the source-books employed by Marx did not consider zero a finite quantity. Therefore this passage states that however small the difference, $x_1 - x = h$ becomes, it always remains different from zero.

96 Here Marx writes simply $x + \dot{x}$ instead of $x + \tau\dot{x}$. Concerning the origins of such replacement, see pp.78-79 of this edition as well as note 60.

97 These notes represent the contents of sheets a to g. Sheets h to n, containing only first-draft fragments or unfinished notes the sense of which is hard to make out, are not published here; concerning them see the Description, pp.468-470 [Yanovskaya, 1968]. Sheets a to g are devoted to an analysis of d'Alembert's method applied to the same example of a compound function which Marx considers in the manuscript 'On the Differential'.

98 The symbols $f(x)$, $f(u)$ are employed here as contractions for the expressions, 'some function in x' and 'some (other) function in u'. In the manuscript 'On the Differential' written later, Marx already designates these functions with different letters in the analysis of the same example.

Additional material on Marx's Mathematical Manuscripts

E. Kol'man

Karl Marx and Mathematics: on the 'Mathematical Manuscripts' of Marx*

The creation of the scientific theory of the revolutionary struggle of the international proletariat to overthrow the capitalist system and to construct socialism made it necessary, as Marx himself indicated, to examine social conditions from the point of view of materialism and dialectics. These must be deduced from the entire complex of real phenomena and verified by the manifold totality, both of the facts of history and of the reality of nature, society and human thought. Thus, one of the necessary prerequisites for the creation of scientific communism was the mastery of the sciences which study the governing laws of the development of nature, the mastery of their results and methods. At the same time the study of the natural sciences, and mathematics as well, from the point of view of their history and interaction with the economic development of society, was necessary for the practical activity of the proletariat as a class coming to power in order consciously to transform society.

With respect to mathematics, dialectical materialism had to solve two closely interrelated problems. On the one hand, it was necessary to generalise the results of mathematics philosophically, and to incorporate them in the scientific world view, the world view of dialectical materialism. On the other hand, the method of materialist dialectics should be used to illuminate the crucial problems of mathematics, thereby enriching the dialectic method. In large measure this work fell to the share of F[riedrich] Engels, since Marx was almost completely occupied with the validation of the laws of the economic development of capitalism and with the practical guidance of the international workers' movement. In spite of this Marx persistently kept track of the progress of natural sciences and the technical

* Translation of 'K.Marks i Matematika (O 'Matematicheskikh rukopisyakh' K. Marksa)', *Voprosy istorii estestvoznaniya i tekhniki*, 1968, No.25, pp.101-112.

achievements of his times, and for almost thirty years, from the late [18]50s right up to his death, was occupied with mathematics a great deal.

These studies were reflected in a number of observations scattered throughout the works of Marx, both on the influence of mathematics on philosophy and on the philosophic elucidation of specific problems of mathematics. In addition, they were expressed in his wide-ranging correspondence, particularly with Engels. Then they were used by Marx in the preparation of his most important work, *Capital*. Finally, the results of his studies were preserved in the extensive manuscripts left behind on Marx's death. These papers were devoted to various problems of mathematics and its history, primarily the problem of the logical and philosophic basis of the differential calculus.

Marx had two motives for his mathematical studies: political economy and philosophy.

Although Marx repeatedly emphasised the specific nature and extraordinary complexity of economic phenomena and the impossibility of comparing them to biological, still less physical, phenomena, nonetheless he considered the application of mathematics not only possible but indeed necessary for the investigation of the general laws of economics. In *Capital* Marx employed a mathematical form in writing down economic laws, by no means solely for illustration. The analysis of the form of value and money, the composition of capital, the rate of surplus value, the rate of profit, the process of transformation of capital, its circulation and turnover, its reproduction, its accumulation, loan capital and credit, differential rents: — Marx accomplished all of this by employing mathematics. Proceeding by means of the simplest algebraic transformations from one formula to another, he next analysed them, interpreted them economically, and formulated new laws. By just such means, for example, Marx derived the dependence of the rate of profit

$$P = \frac{M}{(C+V)}$$

(where C is constant capital, V is variable capital, and M is surplus value)* on the organic composition of capital

$$O = \frac{C}{V}$$

* Constant capital is capital investment; variable capital is labour wages; surplus value is usually written *S* in English-language economic texts — *Trans.*

so that

$$P = \frac{A}{1 + O}$$

(where $A = \frac{M}{V}$ is the rate of production of surplus value), and established the law of the tendency of the average rate of profit to fall. By the very same means he established the inter-relation between the two sectors of capitalist reproduction: the first sector is the production of the means of production:

$$C_1 + V_1 + M_1 = T_1$$

(where T_1 is the total value of the producers' goods sector), and the second sector is the production of the means of consumption,

$$C_2 + V_2 + M_2 = T_2 ,$$

so that, for simple reproduction,*

$$C_2 = V_1 - M_1 .$$

He discovered thereby the general law of the formation of the costs of production and the economic 'mechanism' inevitably leading, under conditions of premonopolistic capitalism, to strongly periodic economic crises.†

The still unpublished preparatory works to the third volume of *Capital* contain Marx's detailed calculations of the quantity $\frac{A.O}{1+O}$, the difference of the rate of surplus value A and the rate of profit P, where Marx represented its variations in the form of a variety of curves. Since the third volume of *Capital*, which is devoted to the process of capitalist production taken as a whole, is a synthesis of the first volume — the immediate process of the production of capital — and the second volume — the process of transformation of capital — Marx tried in his rough drafts to supplement the complete and comprehensive qualitative picture provided in his previous work with a quantitative picture.

Marx did not bring this work, which even in the case of simple reproduction demands rather complicated, although elementary, computations, to completion. The work, however, correctly posed

* In simple reproduction all the value added to the producers' goods is invested in the machinery to produce consumers' goods — *Trans.*

† The significance of these schema for socialist economic planning is examined in the work of M. Ebeseldt (GDR), 'Marx's Schema of Reproduction and the interpretation of Ambiguous Variables', (in Russian) *Ekonomika i matematicheskie metody*, 1968, Vol,IV, No.4, pp.531-535.

the problem of the distribution of surplus value (in the costs of production) under conditions of large-scale reproduction in both sectors in order to obtain maximum profits and also derived the law of periodic crises. These are problems which can only be solved by means of contemporary methods of linear programming. The mechanism of economic crises, however, can also be studied empirically, a method concerning which Marx wrote to Engels on May 31, 1873:

> 'I have just sent Moore a history which *privatim* had to be smuggled in. But he thinks that the question is unsolvable or at least *pro tempore* unsolvable in view of the many parts in which facts are still to be discovered relating to this question. The matter is as follows: you know tables in which prices, calculated by percent etc. etc. are represented in their growth in the course of a year etc. showing the increases and decreases by zig-zag lines. I have repeatedly attempted, for the analysis of crises, to compute these "ups and downs" as fictional curves, and I thought (and even now I still think this possible with sufficient empirical material) to infer mathematically from this an important law of crises. Moore, as I already said, considers the problem rather impractical, and I have decided for the time being to give it up.'*

The mathematician Samuel Moore, who was Marx's adviser in mathematics, was unfortunately not sufficiently well versed; he was obviously unacquainted with Fourier analysis, that branch of applied mathematics which deals with the detection of latent periodicities in complex oscillatory processes, the foundations of which were already contained in J. Fourier's 1822 work, *Analytic Theory of Heat*.

Since Marx believed, according to Paul Lafargue,† that 'a science is not really developed until it has learned to make use of mathematics', he advanced the thesis of the possibility, indeed the necessity, of the application of the mathematical method to research in the social sciences, in political economy in particular. At the same time this did not mean the replacement of political economy and its general laws and methods by mathematics along the lines of the so-called 'mathematical school' of vulgar political economy, headed in England by W. Jevons and in Italy by V. Pareto and others, which had sprung up in the [18]80s in opposition to the bankrupt 'historical school' but

---

* Karl Marx-Friedrich Engels *Werke* [German edition], Vol.33, Berlin, Dietz, 1966, p.82.

† *Reminiscences of Marx and Engels*, Moscow [1956], p.75.

which, like the latter, also argued for a 'harmony of interests' of all classes of capitalist society. Marx made the following observation, in a letter to Engels on March 6, 1868, regarding one of the representatives of this school, Macleod: '. . . a puffed-up ass, who 1) puts every banal tautology into algebraic form and 2) represents it geometrically.'* Thus, according to Marx, as in any other specialised science so in political economy, mathematics can be a powerful tool for research only within the limits of the validity of the theory of that specialised science. Therefore, as his acquaintance the Russian jurist and publicist M.M. Kovalevskii wrote,† Marx devoted himself to the study of mathematics in order to obtain the ability to apply the mathematical method as well as to examine profoundly the works of the mathematical school.

Marx's considered judgement on one of the most important problems of the foundations of geometry, which he expressed in 'The Theory of Surplus Value', the unfinished 4th volume of *Capital*, in connection with a polemic with [Samuel] Bailey, who had incorrectly used the geometric analogy, may serve as an example of his philosophical conclusions on the questions of mathematics. Marx wrote:

> 'If a thing is distant from another, the distance is in fact a relation between the one thing and the other; but at the same time this distance is something different from this relation between the two things. It is a dimension of space, it is a certain length which may as well express the distance of two other things besides those compared. But this is not all. When we speak of the distance as a relation between two things, we presuppose something "intrinsic", some "property" of the things themselves, which enables them to be distant from each other. What is the distance between the syllable A and the table? The question would be nonsensical. In speaking of the distance of two things, we speak of the difference in space. Thus we suppose both of them to be contained in space, to be points of space. Thus we equalise them as being both existences of space, and only after having them equalised *sub specie spatii* we distinguish them as different points of space. To belong to space is their unity.'§

---

* Karl Marx-Friedrich Engels *Sochineniya* [Russian edition], Moscow, Vol.32, p.33.

† *Reminiscences of Marx and Engels*, p.325.

§ Karl Marx, *Theories of Surplus Value: Volume IV of Capital*, part III, Cohen and Ryazanskaya, trans., London, Lawrence & Wishart, 1972, p.143. Editors Ryazanskaya and Dixon note that 'Marx wrote this paragraph in English'.

Here Marx, while analysing the process of abstraction by means of which the geometric concept of 'distance' or 'length' originates, focuses attention not only on the materialistic origin of this concept, the basis of which lies in the 'characteristic' of two comparable objects, but also on its relative character, on its indissoluble connection with space, understood as a material, really existing entity. And all this was in 1861-1863, during the unbroken predominance in science of the Newtonian world view, some forty years before the appearance of the theory of relativity, in which Einstein boldly took to its logical conclusion the idea that 'length' is not simply a superficial abstract measure of a physical body but an integral characteristic of the spatial relationship of two bodies.

Marx's statement on the statistical nature of economic mechanisms as mechanisms of large-scale processes has an exceptionally great methodological significance for mathematical statistics. These mechanisms express the interactions of individual processes in the laws of probability; they dominate over any variations from the mean. Marx repeatedly returned to this problem. For example, in the *Grundrisse* of 1857-1858 he wrote, in the chapter on money:

> 'The value of commodities as determined by labour time is only their *average value*. This average appears as an external abstraction if it is calculated out as an average figure of an epoch, e.g. a pound of coffee is one shilling if the average price of coffee is taken over, let us say, 25 years; but it is very real if it is at the same time recognised as the driving force and the moving principle of the oscillations which commodity prices run through during a given epoch. This reality is not merely of theoretical importance: it forms the basis of mercantile speculation, whose calculus of probabilities depends both on the median price averages which figure as the centre of oscillation, and on the average peaks and average troughs of oscillation above or below this centre.*

Despite the misconception, current for a long time among the majority of Marxists working in the field of economic statistics, that Marx's statements on stochastic processes apply only to capitalist economics, a misconception based on the non-dialectical representation of the accidental and the necessary as two mutually exclu-

---

* Karl Marx, *Grundrisse: Foundations of the Critique of Political Economy*, trans. M. Nicolaus, Penguin Books, London, p.137.

sive antitheses, these statements of Marx — to be sure, in a new interpretation — have enormous significance for a planned socialist economy, in which, since it is a commodity economy, the law of large numbers never ceases to operate.

Hegel's *Science of Logic*, especially the second section to the first book, 'Quantity', was undoubtedly a philosophical stimulus for Marx's mathematical studies. The article 'Hegel and Mathematics', written by the present author together with S.A Yanovskaya,* cites in this connection the following words of Engels:

> 'I cannot fail to comment on your remarks on the subject of Old Man Hegel, to whom you do not attribute a profound mathematical and scientific education. Hegel knew so much mathematics that not one of his students was capable of publishing the numerous mathematical manuscripts left behind after his death. The only person, so far as I know, sufficiently knowledgeable of mathematics and philosophy to perform such a task — is Marx.'†

In the 'Philosophical Notebooks' V.I. Lenin criticised§ the statements of Hegel on the calculus of infinitesimally small quantities contained in the chapter 'Quantity', specifically, that '. . . the justification [for neglecting higher-order infinitesimals — E.K.] has consisted *only* in the *correctness of the results* ("demonstrated on other grounds") . . . and not in the clearness of the subject . . .', that '. . . a certain inexactitude (conscious) is ignored, nevertheless the result obtained is not approximate but *absolutely* exact,' that 'notwithstanding this, to demand *Rechtfertigung* [justification — *Trans.*] here is "not as superfluous" "as to ask in the case of the nose for a demonstration of the right to use it".'** V.I. Lenin made the following remarks: 'Hegel's answer is complicated, *abstrus*, etc. etc. It is a

* This edition p.235

† Afterword to 2nd German edition of *Capital*

§ V.I. Lenin, *Collected Works*, Vol.38, Moscow, Foreign Languages Publishing House, 1963, pp.117-118.

** Note provided by editor of Lenin text: 'An allusion to the couplet "The Question of Right" from Schiller's satirical poem "The Philosophers", which may be translated as follows:

'Long have I used my nose for a sense of smell,
'Indeed, what right have I to this, pray tell?'

question of *higher* mathematics . . .' 'A most detailed consideration of the differential and integral calculus, with quotations — Newton, Lagrange, Carnot, Euler, Leibnitz etc., etc. — showing how interesting Hegel found this "vanishing" of infinitely small magnitudes, this "intermediate between Being and non-Being". Without studying higher mathematics all this is incomprehensible. Characteristic is the title *Carnot*: "Refléxions sur la Métaphysique du calcul infinitésimal"!!!'

It is undoubtedly true that Marx, who had written in 1873:

> 'The mystification which the dialectic suffered at the hands of Hegel does not obscure the fact that Hegel first gave a comprehensive and conscious representation of its general forms of motion. It is necessary to stand it on its feet, in order to reveal the rational kernel beneath the shell of mystification.'*

having already applied his dialectical materialist method which, in his own words, was not only fundamentally 'different from the Hegelian, but is its direct antithesis', since for Marx 'the ideal is nothing other than the material, perceived in a human head and transformed within it',† was extremely tempted to try to discover the secret which seemed to lie at the basis of differential calculus.

Marx's studies of mathematics were known from his correspondence with Engels, particularly the letters from Marx to Engels of January 11, 1858, May 20 1865, July 6, 1863, and August 25, 1879, the letters from Engels to Marx of August 18, 1881 and November 21, 1882, and Marx's answer of November 22, 1882. They may also be evaluated from references in Engels's preface to the second volume of *Capital*, comments in Engels's *Anti-Dühring*, and in his unfinished manuscript, *The Dialectics of Nature*, published for the first time in 1925 in Moscow in the second book of the [Russian-language] *Archives of Marx and Engels*. The Karl Marx-Friedrich Engels Institute, which was founded in 1920, in carrying out the instructions of V.I. Lenin in his letter of February 2, 1921§ to purchase the manuscripts of Marx and Engels located abroad (or photocopies of them), acquired a great many, including photocopies of Marx's mathematical manuscripts preserved in the archive of the German Social-

* Karl Marx-Friedrich Engels, *Sochineniya*, Vol.23, p.22.

† *Ibid.*

§ Leninskii Sbornik, Moscow, 1942, Vol. 34, pp.401-402.

Democratic Party — 863 closely-written quarter-sheets, apparently incomplete; the missing pages were later added, however, so that the entire collection came to a thousand sheets. To work on them the Institute commissioned the German mathematician E.Gumbel, whom R. Mateika and R.S. Bogdan helped to decipher the extremely difficult text.

In 1927 Gumbel published a report in *Letopisi Marksizma* on the manuscripts,* giving a short description of them. He classified the manuscripts into categories: calculations without any text at all; extracts from works read by Marx; outlines of his own works; and finally, finished original works.

Gumbel correctly noted that Marx's choice of sources seemed to be influenced by Hegel, and he presented a (far from complete) list of mathematical works which Marx had summarised: 13 authors and 18 titles. Of these works, the oldest in time was the *Philosophiae Naturalis Principia Mathematica* of Newton, 1687, and the most recent, the textbooks of T.J. Hall and J.W. Hemmings, 1852. They also included the classical works of d'Alembert, Landen, Lagrange, MacLaurin, Taylor and two other works of Newton, *De Analysi per Aequationes Numero Terminorum Infinitas* and *Analysis per Quantitatum Series, Fluxiones et Differentias*.

The contents of the manuscripts, Gumbel indicated, dealt with arithmetic (for example, the effect of a discount on the rate of exchange, the paying off of a bill of exchange, discounts and rebates, raising to a power and extracting the root of an equation, exercises in taking the logarithm, and so forth), geometry (trigonometry, analytic geometry, conic sections), algebra (the elementary theory of equations, infinite series, the concept of function, Cardan's Rule, progressions, the method of indeterminate coefficients), and differential calculus (differentiation, maxima and minima, the Taylor theorem). He reported that the original works which Marx had completed would be published in the 16th volume of [the Russian edition of] the *Complete Works* of Marx and Engels.

In 1931, with the appointment of the well-known activist of the Bolshevik Party V.V. Adoratskii to be director of the Institute, work on the manuscripts was given a new direction. As head of the Marx Study Centre at the time, I was acquainted with the transcribed

* E. Gumbel, 'On the Mathematical Manuscripts of K. Marx', (in Russian) *Letopisi Marksizma*, Moscow, 1927, Vol.3, pp.56-60.

portion of the manuscripts and with the preparatory work toward their publication, and I was convinced that E. Gumbel was unable to appreciate completely either the importance of their publication or their philosophical and historical-mathematical significance. At my suggestion the board of directors of the Institute enlisted for the work on the manuscripts S.A. Yanovskaya, leading a team which was joined by the mathematicians D.A. Raikov and A.I. Nakhimovskaya.

In London in 1931 the Second International Congress of the History of Science and Technology took place, at which a Soviet delegation took part whose members included the author of these lines. The papers of our delegation came out as a separate book with the title *Science at the Crossroads*.* Among the papers included was my own, entitled: 'A Brief Report on the Unpublished Works of Karl Marx pertaining to Mathematics, the Natural Sciences, Technology and Their Histories'. This report discussed: first, the passages from 27 works of natural science which Marx copied and to which he supplied commentaries: on mechanics, physics, chemistry, geology, biology, as well as on electrical technology, metallurgy, agricultural chemistry, and others; second, his works on technology (primarily dating to 1863), treating the history of mills, the history of looms, the problem of automated production in mechanised factories, the development from tools to machines and from machines to mechanised factories, the effect of the mechanisation and rationalisation of production on the development of the textile industry in England and on the situation of the proletariat in the period 1815-1863, the changes in the social system of production at various stages of technological development, the interaction between labour and science, between city and countryside, and so on; and third — Marx's mathematical manuscripts.

In Zurich in 1932 there convened an International Congress of Mathematicians in which a Soviet delegation took part. At the 'Philosophy and History' section of the congress I made the report, 'A New Foundation of the Differential Calculus by Karl Marx',† which

---

* *Science at the Crossroads*: Papers presented to the International Congress of the History of Science and Technology held in London from June 29th to July 3rd, 1931, by the Delegation of the USSR, Kniga Ltd., Bush House, Aldwych, London WC2, 1931. Republished in 1971.

† E. Kol'man, 'A New Foundation of the Differential Calculus' by Karl Marx', [in German], *Verhandlungen des Internationalen Mathematiker-Kongresses*, Vol.2, Sektions-Verträge, Zurich, 1932, pp.349-351.

discussed one of the works contained in Marx's manuscripts. It was of great interest, both for the history of mathematics and for those dealing with the philosophical problems of the scientific worker, since it contains a sketch of the historical development of the concept of the differential and a statement of Marx's viewpoint on the foundation of analysis. This work is of the third category of the manuscripts, and consists of five chapters: 1. The Derivative and the Differential Coefficient [the at that time so-called ratio, $\frac{dy}{dx}$, 2. The Differential and Differential Calculus, 3. The Historical Development of Differential Calculus, 4. The Theorem of Taylor and MacLaurin, 5. A Critique of Newton's Method of Quadratures.

The first part of the third chapter, which forms the nucleus of the entire work, contains a brief account of the methods of Newton, Leibnitz, d'Alembert and Lagrange. The second part, which summarises the first, consists of three sections with the following contents: 1. Mystical Differential Calculus, 2. Rational Differential Calculus, 3. Purely Algebraic Differential Calculus. In another fragment Marx contrasts his own differential method to the methods of d'Alembert and Lagrange. His method differs from the method of Lagrange because Marx really differentiates, thanks to which differential *symbols* appear, while Lagrange applies differentiation to the algebraic binomial expansion.

It is clear from both fragments that Marx, like Hegel, considered all efforts to provide a purely formal-logical foundation for analysis hopeless, just as the attempts to give, beginning with the graphic method, a purely intuitive-visual foundation to it had been naive. He set himself the task of providing a foundation for analysis dialectically, relying on the unity of the historical and logical aspects.

Marx demonstrated both that the new differential and integral calculus came into existence from elementary mathematics, on its own ground, 'as a specific type of calculation which already operates independently on its own ground,' and that 'the algebraic method therefore inverts itself into its exact opposite, the differential method'. (See p.21 in this edition.) Marx valued highly the work of Lagrange, but he did not consider him — as he was usually considered and as Hegel considered him — a formalist and conventionalist who introduced the basic concepts of analysis into mathematics in a purely superficial and derivative manner. Marx appreciated just the opposite in Lagrange, namely, that he revealed the connection between algebra and analysis, that he showed how analysis develops out of algebra. 'The real and therefore the simplest connection of the new with the old is discovered as soon as this new reaches its completed form, and one may say that differential calculus gained this relation through the theorems of Taylor and MacLaurin.' (See p.113)

At the same time, however, Marx reproached Lagrange for not perceiving the dialectical character of this development, for sticking for too long to the domain of algebra, and for insufficiently appreciating the general laws and methods proper to analysis, so that 'in this regard he should only be used as a starting point'. (See Yanovskaya, 1968, p.417) Thus Marx, like a genuine dialectician, rejected both the purely analytic reduction of the new to the old characteristic of the methodology of the mechanistic materialism of the 18th century, and the purely synthetic introduction of the new from outside so characteristic of Hegel.

Reports and articles concerning Marx's mathematical manuscripts also appeared in 1932 in the journals *Za Marksistsko-Leninskoe Estestvoznanie*, *Vestnik Kommunisticheskoe Akademii*, and *Front Nauki i Tekhniki*.* There was a great deal of interest in the manuscripts among the Soviet, as well as the foreign, learned public. Only in 1933,† however, did it become possible, as a result of the work of the team of scholars mentioned above, to publish the first extracts from the manuscripts, in the journal *Pod Znamenem Marksizma* and simultaneously in the collection *Marksizm i Estestvoznanie*, issued on the 50th anniversary of Marx's death by the Marx-Engels Institute. In both publications, the extracts from the manuscripts were accompanied by the article 'On the Mathematical Manuscripts of K. Marx'§ by the team leader S.A. Yanovskaya. The published extracts are three works of Marx dating from the [18]70s and the beginning of the [18]80s. Marx completely finished and prepared to send to Engels the first two — 'The Derivative and the Symbolic Differential Coefficient' and 'The Differential and Differential Calculus'. The third work, 'A Historical Sketch', is an unfinished draft. From the latter, which includes the sections: 1. Mystical Differential Calculus (that is,

---

* *Za Marksistsko-Leninskoe Estestvoznanie*, 1932, No.5-6, pp.163-168; *Vestnik Kommunisticheskoi Akademii*, 1932, No.9-10, pp.136-138; *Front Nauki i Tekhniki*, 1932, No.10, pp.65-69.

† The original has 1932, an obvious misprint.

§ *Pod znamenem marksizma*, 1933, No.1, pp.14-115; *Marksizm i estestvoznanie*, 1933, pp.136-180.

Newton and Leibnitz), 2. Rational Differential Calculus (that is, d'Alembert) and 3. Purely Algebraic Differential Calculus (that is, Lagrange); we introduce here in the team's translation, section 1, in order to acquaint the reader with Marx's exposition. (pp.91-92)

'*1. Mystical Differential Calculus*. $x_1 = x + \Delta x$ from the beginning changes into $x_1 = x + dx$ or $x + \dot{x}$ [Marx uses both the symbol dx of Leibnitz and the $\dot{x}$ of Newton — E.K.] where *dx is assumed by metaphysical explanation*. First it exists, and then it is explained.

'Then, however, $y_1 = y + dy$ or $y_1 = y + \dot{y}$. From this arbitrary assumption the consequence follows that in the expansion of the binomial $x + \Delta x$ *or* $x + \dot{x}$, the terms in $\dot{x}$ and Δx which are obtained in addition to the first derivative, for instance, must be *juggled away* in order to obtain the correct result, etc. etc. Since the real foundation of the differential calculus proceeds from this last result, namely from the *differentials* which anticipate and are not derived but instead are *assumed* by explanation, then $\frac{dy}{dx}$ or $\frac{\dot{y}}{\dot{x}}$, as well, the symbolic differential coefficient, is *anticipated* by this explanation.

'If the increment of $x = \Delta x$ and the increment of the variable dependent on it $= \Delta y$, then it is self-evident (*versteht sich von selbst*) that $\frac{\Delta y}{\Delta x}$ represents the ratio of the increments of x and y. This implies, however, that Δx figures in the denominator, that is the increase of the independent variable is in the denominator instead of the numerator, not the reverse; while the final result of the development of the differential form, namely the *differential*, is also given in the very beginning by the assumed differentials.*

'If I assume the simplest possible (*allereinfachste*) ratio of the dependent variable y to the independent variable x, then $y = x$. Then I know that $dy = dx$ or $\dot{y} = \dot{x}$. Since, however, I seek the derivative of the independent [variable] x, which here $= \dot{x}$, I therefore have to divide both sides by $\dot{x}$ or dx; so that:

$$\frac{dy}{dx} \text{ or } \frac{\dot{y}}{\dot{x}} = 1$$

'I therefore know once and for all that in the symbolic differential coefficient the increment [of the independent variable] must be placed in the denominator and not in the numerator.

'Beginning, however, with functions of x in the second degree, the *derivative* is found immediately by means of the binomial theorem [which provides an expansion] where it appears ready-made (*fix und fertig*) in the second term combined with dx or $\dot{x}$; that is with the increment of the first degree + the terms to be juggled away. The *sleight of hand* (*Eskamotage*) however, is unwittingly mathematically correct, because it only juggles away errors of calculation arising from the original sleight of hand in the very beginning.

$x_1 = x+\Delta x$ is to be changed to

$$x_1 = x + dx \quad \text{or} \quad x + \dot{x} ,$$

whence this differential binomial may then be treated as are the usual binomials, which from the technical standpoint would be very convenient.

'The only question which still could be raised: why the mysterious suppression of the terms standing in the way? That specifically assumes that one knows they stand in the way and do not truly belong to the derivative.

'The *answer* is very simple: this is found purely by experiment. Not only have the true derivatives been known for a long time, both of many more complicated functions of x as well as of their analytic forms as equations of curves, etc., but they have also been discovered by means of the most decisive experiment possible, namely by the treatment of the simplest algebraic function of second degree, e.g.:

$$y = x^2$$

$$y + dy = (x + dx)^2 = x^2 + 2xdx + dx^2 ,$$

$$y + \dot{y} = (x + \dot{x})^2 = x^2 + 2x\dot{x} + \dot{x}^2 .$$

'If we subtract the original function, $x^2 (y = x^2)$ from both sides, then:

$$dy = 2xdx + dx^2$$

$$\dot{y} = 2x\dot{x} + \dot{x}\dot{x} ;$$

I suppress the last terms on both [right] sides; then:

$$dy = 2x\,dx\ ,\ \dot{y} = 2x\dot{x}\ ,$$

and further

$$\frac{dy}{dx} = 2x\ ,$$

or

$$\frac{\dot{y}}{\dot{x}} = 2x\ .$$

'We know, however, that the first term out of $(x + a)^2$ is x^2; the second $2xa$; if I divide this expression by a, as above $2xdx$ by dx or $2x\dot{x}$ by $\dot{x}$, we then obtain $2x$ as the first derivative of x^2, namely the increase in x, which the binomial has added to x^2. Therefore the dx^2 or $\dot{x}\dot{x}$ had to be suppressed in order to find the derivative; completely neglecting the fact that nothing could begin with dx^2 or $\dot{x}\dot{x}$ in themselves.

'In the experimental method, therefore, one comes — right at the second step — necessarily to the insight that dx^2 or $\dot{x}\dot{x}$ has to be juggled away, not only to obtain the true result but any result at all.

'Secondly, however, we had in

$$2x\,dx + dx^2 \quad \text{or} \quad 2x\dot{x} + \dot{x}\dot{x}$$

the true mathematical expression (second and third terms) of the binomial $(x + dx)^2$ or $(x + \dot{x})^2$. That *this mathematically correct result* rests on *the mathematically basically false assumption* that $x_1 - x = \Delta x$ is from the beginning $x_1 - x = dx$ or $\dot{x}$, was not known.

'In other words, instead of using sleight of hand, one obtained the same result by means of an algebraic operation of the simplest kind and presented it to the mathematical world.

'Therefore, mathematicians (*man . . . selbst*) really believed in the mysterious character of the newly-discovered means of calculation which led to the correct (and, particularly in the geometric application, surprising) results by means of a positively false mathematical procedure. In this manner they became themselves mystified, rated the new discovery all the more highly, enraged all the more greatly the crowd of old orthodox mathematicians, and elicited the shrieks of hostility which echoed even in the world of non-specialists and which were necessary for the blazing of this new path.'

In an analogous manner Marx critically analysed both the method of d'Alembert as well as that of Lagrange and, as already mentioned, opposed all three methods with his own. It consists of first forming, for $y = f(x)$, the 'preliminary derivative',

$$\varphi(x_1,x) = \frac{f(x_1) - f(x)}{x_1 - x}$$

which is assumed to be continuous at $x_1 - x$ and whose value at $x_1 = x$ is equal to $f'(x)$. In the case of the power function $y = x^n$, the ratio $(x_1^n - x^n)/(x_1 - x)$ is transformed into the polynomial $x_1^{n-1} + xx_1^{n-2} + \ldots + x^{n-2}x_1 + x^{n-1}$, which for $x_1 =$ gives $f'(x) = nx^{n-1}$. Marx then introduces the symbolic representation of this process, by which the 'preliminary derivative' $\frac{\Delta y}{\Delta x}$ is reduced to $f'(x) = \frac{dy}{dy}$, where the symbolic differential coefficient $\frac{dy}{dx}$ has an immediate meaning only as a unit (and not as the two partial quantities dy and dx). However, notes Marx, since the equality

$$dy = f'(x)dx \qquad (*)$$

is mathematically correct and is not reduced to the tautology

$$0 = 0$$

it therefore is an *operative* formula [emphasis in original — *Trans.*], applicable to complicated functions, making it possible to reduce an entire differentiation of its constituent functions. In this way, he points out, we obtain the dialectical *reversal* of the method: we now proceed not only from the real mathematical process of the formation of the derivative to its symbolic expression, but rather on the contrary, operating on the symbolic formula (*) and forming the ratio $\frac{dy}{dx}$ we arrive at the expression of the derivative of the function. Consequently Marx, having not only discovered that the differential is the major linear portion of the increment but is also an operative symbol, proceeded along a path which we today would call *algorithmic*, in the sense that it consists of a search for an exact instruction for the solution, by means of a finite number of steps, of a certain class of problems. He was on a path which has been the fundamental path of the development of mathematics. Thanks to the dialectical materialist method which in his hands was a powerful, effective tool of research, Marx was able, without being a mathematician, to reveal the property of the differential used as an operational symbol, thus anticipating, as the Soviet mathematician V.I. Glivenko has shown, the idea of the eminent French mathematician G. Hadamard, enunciated in 1911 in connection with the application of this concept of functional analysis.*

Despite the philosophical and historical significance of the foundation of differential calculus provided by Marx, it did not enter into mathematics, which developed another path unknown to him. The sources which he studied (and their number was significantly greater than Gumbel reported in his article, which did not mention even those textbooks of analysis, such as those of J.-L. Boucharlat and J. Hind, which Marx outlined in detail) made no mention of the works of A. Cauchy (*Cours d'analyse* and *Résumé des leçons sur le calcul infinitésimal*) in which in 1821-1823 he developed the theory of limits, a theory which, although it contained shortcomings which were later (1880) cleared up by K. Weierstrass, nonetheless incorporated a great deal of rigour and rendered the foundation proposed by Marx superfluous, although it did not diminish its historical and philosophical value. Marx did not know and could not have known of the work of the outstanding logician, mathematician and philosopher of Prague, B. Bolzano, who in 1816-1817 defined the concepts of limit, continuity, the convergence of series, and others — concepts which laid the basis of present-day analysis — since these works as well as others of 1830-1848 which contained the beginnings of set theory and the theory of real numbers remained unknown for a long time. Only a hundred years later did they become the property of mathematicians. Naturally, Marx did not consider, therefore, the problems of continuity, the differentiability of functions, the axiomatisation of analysis, and so on.

The value of Marx's mathematical manuscripts, however, is by no means restricted to his method providing a foundation for differential calculus and his critique of preceding methods. The complete significance of the manuscripts was only revealed when they were all deciphered and scientifically systematised. Beginning with 1932 and with the publication in 1933 of the three works mentioned from the deciphered manuscripts (which Gumbel had not given the attention they deserved), the Swedish mathematician Wildhaber first began working on behalf of the Marx-Engels Institute. Work on the manuscripts was resumed in the 1950s, and somewhat later (1960-1962) G.F. Rybkin became interested. All this work — deciphering, translation, research, and compilation of sources — was conducted under the leadership of S.A. Yanovskaya, who, despite an extraordinary

* V.I. Glivenko, 'The Concept of the Differential in Marx and Hadamard' (in Russian *Pod Znamenem Marksizma*, 1934, No.5, pp.79-85.

load of teaching and preparing graduate students, despite a painful illness, gave the enterprise all of her energy and her enormous knowledge of the history of mathematics and its philosophical problems, transforming it into her life's work. S.A. Yanovskaya's commentaries on the manuscripts (both the one cited above and those contained in the volume prepared by the Institute of Marxism-Leninism of the Central Committee of the CPSU) by themselves constitute an important scientific work. One of her many students, K.A. Rybnikov, performed significant work in the preparation of the manuscripts for publication (in particular, the difficult research and collation of sources). The volume was prepared for publication by the historian O.K. Senekina, member of the Institute of Marxism-Leninism, and the mathematician A.Z. Rybkin, editor of the Nauka press.

As a result of all this work lasting many years (S.A. Yanovskaya laboured on the manuscripts until her death in October 1966), a book has appeared which contains Marx's ideas on a series of the most important problems in the history of mathematics as a whole and of its individual concepts, as well as on their epistemological [original: 'gnoseological' — *Trans.*] significance, ideas which, despite the head-spinning pace of the development of mathematics in the '80s of the last century — among which and in particular including its logical-philosophical basis — have not lost their contemporaneity in the slightest. For historians of mathematics and for philosophers working with the philosophical problems of mathematics, Marx's views will serve as a guide — not in the form of a quotation, every letter of which is followed as if counting out an emergency ration, but rather in the form of a matchless example of creative, concrete application of dialectical thinking.

In addition, the mathematical manuscripts of Marx once again confirm the truth of the words Engels spoke at the graveside of his great friend. Speaking of Marx as the scientist who had discovered the law of the development of human history and the law of motion of capitalist production, Engels said: 'Two such discoveries would be enough for one lifetime. Happy the man to whom it is granted to make even one such discovery. But in every single field which Marx investigated — and he investigated very many fields, none of them superficially — in every field, even in that of mathematics, he made independent discoveries.'*

* Quoted from Marx-Engels *Selected Works*, Volume Two, p.153-154, Foreign Language Publishing House, Moscow. The speech was re-translated into English from the only written version, in the German-language *Sozialdemokrat*, Zurich, March 22 1883.

Chronology

CHRONOLOGY OF MARX'S MATHEMATICAL MANUSCRIPTS

An exact chronology of manuscripts is not now possible, but approximate dates have been attributed to each of the mathematical manuscripts by researchers at the Marx-Engels Institute in Moscow. The following list of manuscripts, with archival manuscript numbers and attributed dates, is taken from the exhaustive Description which comprises the second half of the volume being translated. All manuscripts described there are included.

Archival No.	Topic (Date)
147	Algebraic power series (1846)
210	Simultaneous equations; binomial expansion (1846)
497	History of mathematics (Sept.-Oct. 1851)
1052	Arithmetic (1857-1858)
1153	Geometrical problems (no date)
1922	Problem of the tangent to the parabola (1865-1866)
2055	History of mathematics (1867-1869)
2388	Commercial arithmetic (1869)
2400	Commercial arithmetic (1869)
2759	Summary of trigonometry (no date)
2760	Theory of conic sections (no date)
2761	Theory of conic sections (no date)
2762	Theory of conic sections (no date)
2763	Differential calculus (1872-1881?)
3704	Differential calculus (no date)
3881	Commercial arithmetic (March 1878)
3888	Differential calculus (1878)
3931	Commercial arithmetic (no date)
3932	Theory of equations (1875?-1880?)

Archival No.	Topic (Date)
3933	Assorted algebraic topics (after 3932)
3934	Assorted algebraic topics (after 3932)
3935	Assorted algebraic topics (after 3932)
3999	Successive differentiation (1875?-1880?)
4000	Taylor's and MacLaurin's Theorems (no date)
4001	Taylor's and MacLaurin's Theorems; Lagrange's Theory (after 4000)
4002	Taylor's Theorem (after 4001)
4003	Lagrange's Theory of Derived Functions (no date)
4036	Differential calculus (1880s)
4037	Differential calculus (after 4036)
4038	First draft of 'On the Differential'; historical sketch
4039	Finite differences (1870s)
4040	Table of the formulae of calculus (no date)
4048	Various mathematical topics (no date)
4143	Analysis of d'Alembert's method (no date)
4144	Comparison of d'Alembert's method to the algebraic and On the ambiguity of the term 'limit' (no date)
4145	Comparison of Marx's method to d'Alembert's (no date)
4146	Draft of 'On the Concept of the Derived Function' (no date)
4147	'On the Concept of the Derived Function' (1881)
4148	Second, third drafts of 'On the Differential' (no date)
4149	Fourth draft (supplement) of 'On the Differential' (no date)
4150	'On the Differential' (1881)
4300	Lagrange's method (no date)
4301	Taylor's theorem according to Hall and Boucharlat (no date)
4302	Taylor's Theorem (1882)

Index

INDEX OF SOURCES CONSULTED BY MARX

Page numbers in italics refer to appendices

BOUCHARLAT, J.-L., 'Elementary Treatise on the Differential and Integral Calculus', English version of third French edition, Cambridge-London (1828), 24, 65, *143*, *149-151*, *173-181*, *186-190*

D'ALEMBERT, J., 'Traité de l'équilibre et du mouvement des fluides', Paris (1754), 75, 76, 97

EULER L., 'Introductio in analysi infinitorum', Lausanne (1748) 75

EULER L., 'Institutiones calculi differentialis cum ejus usu in analysi finitorum ac doctrina serierum', Berlin (1755), 75, *160-164*, *182*, *185*

HIND L., 'The principles of the differential calculus; with its application to curves and curve surfaces', second ed., Cambridge (1831), 99, *143-154*, *165*, *185*

LACROIX, S.F., 'Traité du calcul différentiel et du calcul intégral', 3 Vol., second edition, Paris (1810-1819), *145*, *153*, *154*, *171*

LACROIX, S.F., 'Complément des éléments d'algèbre', 7th edition, Paris (1863), *171*

LAGRANGE, J.L., 'Théorie des fonctions analytiques', Paris, 1813, Oeuvres Lagrange, Vol.IX, Paris (1881), 75, 76, 99, *154-155*, *172*

LANDEN, J., 'A discourse of the residual analysis', London (1758), *172*

LANDEN, J., 'The residual analysis', London (1764), 113, *155*, *165-172*

LHUILIER, S., 'Principiorum calculi differentialis et integralis exposito elementaris', Tübingen (1795), *182*

MACLAURIN, C., 'A treatise of algebra in 3 parts', 1st ed. (1748), 6th ed., London (1796), *186*

MOIGNO, F., 'Leçons de calcul différentiel et de calcul intégral, rédigées d'après les méthodes et les ouvrages publiés ou inédits de M. L.A. Cauchy', 2 Vol., Paris (1840 and 1844), 75

NEWTON, I., 'Philosophiae naturalis principia mathematica', London (1687), 75, 76, *156-159*

NEWTON, I., 'Arithmetica universalis, sive de compositione et resolutione arithmetica liber', Cambridge (1707), 112

NEWTON, I., 'Analysis per quantitatum series fluxiones et differentias, cum enumeratione linearum tertii ordinis' (1711), 75, 76

TAYLOR, B., 'Methodus incrementorum directa et inversa', London (1715), 75, *182*

INDEX OF SUBJECTS